LONDON MATHEMATICAL SOCIETY LECTURE NOTE SERIES

Editor: PROFESSOR G. C. SHEPHARD, University of East Anglia

Already published in this series

1. General cohomology theory and K-theory, PETER HILTON.
2. Numerical ranges of operators on normed spaces and of elements of normed algebras, F. F. BONSALL and J. DUNCAN.
3. Convex polytopes and the upper bound conjecture, P. McMULLEN and G. C. SHEPHARD.
4. Algebraic topology: A student's guide, J. F. ADAMS.
5. Commutative algebra, J. T. KNIGHT.
6. Finite groups of automorphisms, NORMAN BIGGS.
7. Introduction to combinatory logic, J. R. HINDLEY, B. LERCHER and J. P. SELDIN.
8. Integration and harmonic analysis on compact groups, R. E. EDWARDS.
9. Elliptic functions and elliptic curves, PATRICK DU VAL.
10. Numerical ranges II, F. F. BONSALL and J. DUNCAN.
11. New developments in topology, G. SEGAL (ed.).
12. Symposium on complex analysis Canterbury, 1973, J. CLUNIE and W. K. HAYMAN (eds).
13. Combinatorics, Proceedings of the British combinatorial conference 1973, T. P. McDONOUGH and V. C. MAVRON (eds).
14. Analytic theory of abelian varieties, H. P. F. SWINNERTON-DYER.
15. An introduction to topological groups, P. J. HIGGINS.
16. Topics in finite groups, TERENCE M. GAGEN.
17. Differentiable germs and catastrophies, THEODOR BRÖCKER and L. LANDER.
18. A geometric approach to homology theory, S. BUONCRISTIANO, C. P. ROURKE and B. J. SANDERSON.
19. Graph theory, coding theory and block designs, P. J. CAMERON and J. H. VAN LINT.
20. Sheaf theory, B. R. TENNISON.
21. Automatic continuity of linear operators, ALLAN M. SINCLAIR.
22. Presentations of groups, D. L. JOHNSON.
23. Parallelisms of complete designs, PETER J. CAMERON.
24. The topology of Stiefel manifolds, I. M. JAMES.
25. Lie groups and compact groups, J. F. PRICE.

"I WISH YOU'D NEVER GONE TO THE CONFERENCE ON TRANSFORMATIONS"

London Mathematical Society Lecture Note Series. 26

Transformation Groups

Proceedings of the Conference in the University of Newcastle upon Tyne, August 1976

Edited by CZES KOSNIOWSKI

CAMBRIDGE UNIVERSITY PRESS
CAMBRIDGE
LONDON NEW YORK MELBOURNE

Published by the Syndics of the Cambridge University Press
The Pitt Building, Trumpington Street, Cambridge CB2 1RP
Bentley House, 200 Euston Road, London NW1 2DB
32 East 57th Street, New York, NY 10022, USA
296 Beaconsfield Parade, Middle Park, Melbourne 3206, Australia

ISBN 0 521 21509 9

First published 1977

Printed in Great Britain
at the University Printing House, Cambridge
(Harry Myers, University Printer)

CONTENTS

PART ONE

PART TWO (SUMMARIES AND SURVEYS)

PREFACE

In August 1976 a conference on Transformation Groups was held at the University of Newcastle upon Tyne with participants from the following countries: Canada, Eire, Finland, France, Japan, Norway, Poland, United Kingdom, United States of America, West Germany.

Manuscripts were received from all the speakers at the conference. Other papers were also submitted, four of which were accepted for these proceedings. The articles have been divided into two parts (with summaries and surveys appearing in the second part).

ACKNOWLEDGEMENTS

I would like to to thank very much indeed the following:

The London Mathematical Society and the University of Newcastle upon Tyne for financial assistance towards the conference.

The secretaries of the School of Mathematics, University of Newcastle upon Tyne, for typing the papers. The majority of the typing was by Pauline Harley and Janet McKay. Additional typing was by Joyce Edger and Carol Reynolds

The referees of papers - their assistance and advice was invaluable.

CZES KOSNIOWSKI

NEWCASTLE UPON TYNE

SEPTEMBER 1976

PART ONE

GENERATORS AND RELATIONS FOR GROUPS OF HOMEOMORPHISMS

HERBERT ABELS

The aim of the present paper is to unify and generalize the proofs of results of Behr, Gerstenhaber and Macbeath concerning the theme of the title.

I. RESULTS

1.1 NOTATIONS. Let the group G act on the topological space X, i.e. suppose a homeomorphism of G into the group of homeomorphisms of X is given. By gx we denote the image of the point $x \in X$ under the homeomorphism corresponding to $g \in G$. For $M \subset G$, $A \subset X$ let $MA = \{gx;\ g \in M,\ x \in A\}$.

For any two subsets A, B of X define

(1.1.1) $$G(A,B) := \{g \in G;\ gA \cap B \neq \emptyset\}.$$

Obviously

(1.1.2) $$G(B,A) = [G(A,B)]^{-1}$$

(1.1.3) $$G(A,gB) = g\cdot G(A,B)$$

(1.1.4) $$G(gA,B) = G(A,B)\cdot g^{-1}$$

Let F be a non empty subset of X (think of F as a "fundamental set"). Define

(1.1.5) $$E := G(F,F)$$

(think of E "Erzeugendenmenge"). Let H be a group, $q{:}H \to G$ be a homomorphism. Suppose a section $s{:}E \to H$ is given, i.e. a map $s{:}E \to H$

such that $q \circ s = id|_E$.

In applications H will always be a group with generators E and certain relations, which hold in G, $q:H \to G$ will be the homomorphism induced by the inclusion $E \to G$, the section s is the obvious one. The problem is: Under which conditions is q an isomorphism? We express the relations that hold in H by multiplicative properties of the section s.

1.2 HYPOTHESES

(1.2.1) (Multiplicative hypothesis) If $g_1F \cap g_2F \cap F \neq \emptyset$ we have

$$s(g_1^{-1})\ s(g_2) = s(g_1^{-1}g_2).$$

Both sides of this equality are defined because $E = E^{-1}$ by 1.1.2 and $g_1^{-1}g_2 \in G(F,F) = E$.

Further hypotheses are:

(1.2.2) $GF = X$

(1.2.3) $s(E)$ generates H

(1.2.4) F is connected

(1.2.5) X is connected and simply connected.

1.3 RESULTS

THEOREM 1. $q:H \to G$ is an isomorphism if (1.2.1) through (1.2.5) hold and $G\overset{\circ}{F} = X$ (e.g. if F is open).

THEOREM 2. $q:H \to G$ is an isomorphism if (1.2.1) through (1.2.5) hold, F is closed in X and $\{gF;\ g \in G\}$ is a locally finite cover of X.

Macbeath [6] proved Theorem 1 for open F, Theorem 2 for groups of isometries. The case of finite E in Theorem 2 was proved by Behr [1], with a bigger set of relations. Cf also [5]. Swan

[11] considers the case F open, $\pi_o(F) = \pi_o(X) = 0$ and gives a detailed description of ker(q) if $\pi_1(X) \neq 0$.

We actually prove a common generalization of Theorems 1 and 2 (Theorem 4.5). A result of Soulé [10] is an easy application (see 4.6).

For our next result we need to following definition. The cover $\{gF;\ g \in G\}$ of X is called G-numerable if there is a partition of unity $\{p_g;\ g \in G\}$ with $\mathrm{supp}(p_g) \subset gF$ such that $p_g(gx) = p_e(x)$ for every $x \in X$, $g \in G$. For example, if X is normal, F is open and $\{gF;\ g \in G\}$ is a locally finite cover of X, or if X is normal and F is a neighbourhood of a closed subset of F' and $\{gF';\ g \in G\}$ is a locally finite cover of X, then $\{gF;\ g \in G\}$ is a G-numerable cover of X.

THEOREM 3. $q: H \to G$ is an isomorphism if (1.2.1) and (1.2.3) hold, $\{gF;\ g \in G\}$ is a G-numerable cover of X and $\pi_o(F) = \pi_o(X) = \pi_1(X) = 0$.

1.4 IDEA OF PROOF. It is easy to prove that q is surjective (see Section II). If in Theorems 1 and 2 we drop the assumption that X be simply connected, we obtain a covering space $Y \xrightarrow{p} X$ (i.e. a locally trivial sheaf) with the following properties:

(1) H acts on Y and p is an H-map, i.e. $p(hy) = q(h)p(y)$.

(2) ker q acts as a group of covering transformations of p. The action is free and transitive on the fibres of p.

(3) There is a section for p over F.

Hence if any such covering space of X is trivial, $q: H \to G$ is an isomorphism, in particular if X is simply connected.

The main difficulty of the proof is to define a topology on Y.

The proof of Theorem 3 makes use of the nerve of the covering and its geometric realization. It makes use of fundamental groups instead of covering spaces. There is a similar generalization as above (see Theorem 5.4).

I thank H. Behr for helpful conversations. This paper actually grew out of a talk in a seminar of Behr's. I also thank the referee for useful hints to the literature.

II SURJECTIVITY OF q

Notations as in 1.1. The following result is well known [9, no.9].

2.1. THEOREM. *Suppose* $GF = X$, EF *is a neighbourhood of* F *and* X *is connected*. *Then* E *generates* G.

PROOF. Let G_o be the subgroup of G generated by E. The set $\{X_i = g_iG_oF;\ g_iG_o \in G/G_o\}$ is a cover of X, since $GF = X$. The X_i's are disjoint: $g_1G_oF \cap g_2G_oF \neq \emptyset$ implies that $G_og_2^{-1}g_1G_o$ contains an element of $E \subset G_o$, so $g_2^{-1}g_1 \in G$, hence $g_1G_o = g_2G_o$. Since EF is a neighbourhood of F, each $X_i = g_iG_oF = g_iG_o(EF)$ is a neighbourhood of itself, i.e. open. So the $X_i = g_iG_oF$ form an open disjoint cover of X. If X is connected, there is only one of them: $\# G/G_o = 1$, so $G = G_o$.

III THE SET Y

For the whole Section III we use the notations of 1.1 and assume only the multiplicative hypothesis (1.2.1): If $g_1F \cap g_2F \cap F \neq \emptyset$ we have

$$(3.1) \qquad s(g_1^{-1})\, s(g_2) = s(g_1^{-1} g_2).$$

This implies for $g_1 = g_2 = e$ the neutral element

$$(3.2) \qquad s(e) = e.$$

For $g_1 \, g_2 = e$ we obtain

$$(3,3) \qquad s(g^{-1}) = (s(g))^{-1} \qquad \text{for } g \in E,$$

which we sometimes denote by $s(g)^{-1}$.

We have an action of H on X defined by

$$(3.4) \qquad hx := q(h)x, \qquad h \in H,\ x \in X.$$

Define

$$(3.5) \qquad Z := \{(x,h) \in X \times H;\, h^{-1}x \in F\}.$$

The relation on Z

$$(3.6) \quad (x_1,h_1) \sim (x_2,h_2) \quad \text{if and only if} \quad x_1 = x_2 \text{ and } h_1^{-1}\, h_2 \in s(E)$$

is an equivalence relation by our multiplicative hypotheis (3.1). We define

$$(3.7) \qquad Y = Z/\sim.$$

The main point of the proof will be to endow Y with a suitable topology. We need some preparations. We denote the equivalence class of $(x,h) \in Z$ by $[x,h]$. We have an action of H on Y defined by $h_1[x,h] = [h_1x, h_1\, h]$. The projection $p: Y \to X$, $p([x,h]) = x$ is an H-map. We have a section $t: F \to Y$, namely

$$(3.8) \qquad t(x) = [x,e] \quad \text{for} \quad x \in F.$$

Our definitions imply

(3.9) $\qquad H\, t(F) = Y.$

(3.10) $\qquad H(t(x), t(F)) = s(G(x,F))$ for $x \in F$,

hence

(3.11) $\qquad H(t(F),\ t(F)) = s(E).$

The homomorphism $q{:}H \to G$ is to be analysed. Set $K = \ker q$. The next lemma shows that K is a good candidate for the group of covering transformations of p.

3.12 LEMMA. K _acts freely on_ Y _and simply transitively on the non-empty fibres_ $p^{-1}(x)$ _of_ p.

PROOF. We have to show first that $K_y = \{k \in K;\ ky = y\}$ contains only the neutral element. By (3.9) it suffices to prove that claim for $y \in t(F)$, say $y = [x,e]$. If $ky = [kx,k] = [x,e]$, we have $k \in s(E) \cap K$. But $s(E) \cap K = \{e\}$, since $q|\,s(E){:}\ s(E) \to E$ is bijective with inverse mapping $g \to s(g)$.

K acts on the fibres of p. It remains to prove that K acts transitively on the non empty fibres $p^{-1}p(y)$ of p, $y \in Y$. Again, we may assume, $y = [x,e]$. Suppose $z \in p^{-1}p(y)$. By (3.9) there is an element $h \in H$ and a point $[x_1,e] \in t(F)$ such that $z = h\,[x_1,e] = [hx_1,g]$. Since $p(y) = p(z)$ we have $x = hx_1 = q(h)x_1$, so $q(h) \in E$. For $h_1 = s(q(h)) \in s(E)$ we have $h_1[x_1,e] = [x,e] = y$. Hence $z = h[x_1,e] = h\cdot h_1^{-1}$ and $k = h\cdot h_1^{-1} \in K$.

Note that k is the unique element of K such that $ky = z$. In particular: $hs(q(h))^{-1}$ is the same element of K for every $h \in H$ such that $z = [x,h]$. So the proof actually yields the inverse mapping of

$$K \times F \to p^{-1}(F)$$

(3.13) $$(k,x) \to kt(x) = [x,k]$$

namely $$(h\, s(q(h))^{-1},\ x) \leftarrow [x,h].$$

The next lemma makes explicit the properties of the topology of Y we want.

3.14 LEMMA. *Suppose* Y *is endowed with a topology such that*

(a) *Every* $h \in H$ *acts as a homeomorphism on* Y.

(b) $p:Y \to X$ *is a sheaf*. (*i.e. a local homeomorphism, i.e. every point* $y \in Y$ *has an open neighbourhood* U *such that* $p|U : U \to p(U)$ *is a homeomorphism and* p(U) *is open in* X_1).

Then $p:Y \to X$ *is a covering, i.e. a locally trivial sheaf*, K *acts as a group of covering transformations of* p, *transitively on the non empty fibres of* p.

PROOF. Suppose U as in (b). Then $p^{-1}(p(U)) = \bigcup_{k \in K} kU$ is the disjoint union of the open sets kU. Endow K with the discrete topology. In the commutative diagram

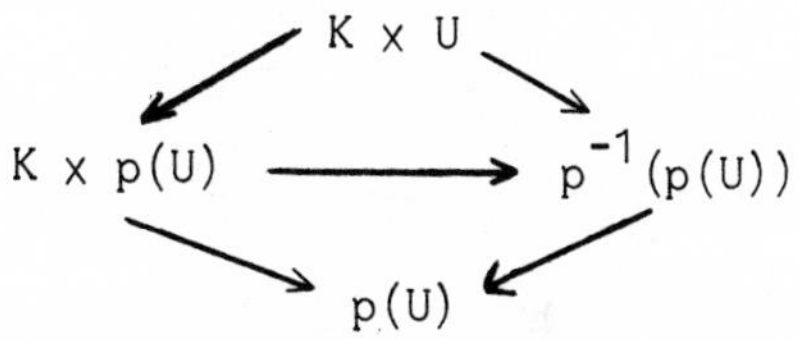

the upper diagonal maps are homeomorphisms, hence so is the horizontal map, yielding the local triviality of the sheaf $p:Y \to X$.

We need a more technical version of 3.14. We call a map $r:A \to Y$ a *section* (for p) if $p \circ r = id_A$. Note that Y is not supposed to have a topology yet. If $r:A \to Y$ is a section, so is ${}_h r:hA \to Y$

defined by

$$(3.15) \qquad {}_h r(hx) = hr(x).$$

As usual, two maps defined in neighbourhoods of the same point $x \in X$ are said to have the same germ at $x \in X$, if and only if they coincide in some neighbourhood of x.

3.16 LEMMA. Suppose we are given for every point $x \in X$ an open neighbourhood U_x of x and a section $r_x : E_x \to Y$ with the following properties:

(a) $$r_x | U_x \cap F = t | U_x \cap F$$

(b) If $x_1 \in F$, $x_2 \in F$, $x_2 = hx_1$, $h \in s(E)$, then r_{x_2} and ${}_h r_{x_1}$ have the same germ at x_2.

(c) Let $x_1 \in F$, $x \in U_{x_1}$. There is an $h \in H$ such that $hx = x_2 \in F$ and r_{x_2} and ${}_h r_{x_1}$ have the same germ at x_2.

Then Y has a unique topology satisfying 3.14 and such that the sections r_x are continuous. In particular $t : F \to Y$ is a continuous section.

PROOF. The proof is given in the language of presheaves. One could give it also by defining neighbourhood bases of the points of Y. Let U be an open subset of X. Define $R(U)$ to be the set of sections $r : U \to Y$ with the following property: For any $x \in U$ there is an $h \in H$ and an $x_1 \in F$ such that $hx_1 = x$ and r and ${}_h r_{x_1}$ have the same germ at x. The $R(U)$ obviously form a presheaf, satisfying the two Serre conditions.

Furthermore,

(i) If $r \in R(U)$ then ${}_h r \in R(hU)$

(ii) $r_x \in R(U_x)$ for $x \in F$

(iii) $r_x(x) = t(x)$ for $x \in F$

(iv) For any point $y \in Y$ there is an open neighbourhood U of $p(y) = x$ and an $r \in R(U)$, such that $r(x) = y$.

(v) The set $\{x \in U;\ r_1(x) = r_2(x)\}$ of points where two sections r_1, r_2 of R(U) coincide, is open.

(i) is immediate from the definition of R(U); (ii) it follows from hypothesis (c); (iii) from (a); (iv) from (i), (iii) and 3.9. Finally (v) follows from hypothesis (b): Note that $h_1 x_1 = x = h_2 x_2$, ${}_{h_i}r_{x_i}(x) = r(x)$ for $i = 1,2$, or more explicitly ${}_{h_i}r_{x_i}(x) = h_i \cdot r_{x_i}(x_i) = h_i t(x_i) = [h_i x_i, h_i] = [x, h_i] = r(x)$ implies $h_1^{-1} \cdot h_2 \in s(E)$ by 3.6, so hypothesis (b) can be applied for $h = (h_1^{-1} \cdot h_2)^{-1} \in s(E)$ by 3.3.

Now let $\mathfrak{R}$ be the sheaf associated to the presheaf R. Then the evaluation map $\mathfrak{R} \to Y$ is bijective by (iv) and (v). If Y is endowed with the transported topology, 3.14 is satisfied, (a) by (i). Since the presheaf R satisfies the Serre conditions, R(U) is the set of continuous sections $U \to Y$. In particular $r_x : U_x \to Y$ is continuous by (ii). The uniqueness of the topology of Y with these properties is obvious. Finally, $t : F \to Y$ is continuous by hypothesis (a).

IV PROOF OF THEOREMS 1 AND 2

We use the notations of section III and the multiplicative hypothesis (1.2.1) = (3.1). Define for $x \in X$

(4.1) $$\gamma(x) = G(F, x) = \{g \in G;\ x \in gF\}$$

Note:

(4.2) $$\gamma(gx) = g\,\gamma(x)$$

by (1.1.3).

The hypothesis of the following lemma generalizes the extra hypotheses of both Theorems 1 and 2 (see 4.4).

4.3 LEMMA. Assume that every point $x \in F$ has a neighbourhood U_x such that $(\gamma(x) \cap \gamma(y))F$ is a neighbourhood of y for every $y \in U_x$. Then there is a set of sections r_x as in 3.16.

PROOF. We may assume that U_x is open. Define for $z \in U_x$:

$$r_x(z) = [z, s(g)] \quad \text{for} \quad g \in \gamma(x) \cap \gamma(z).$$

By our assumptions the set $\gamma(x) \cap \gamma(z)$ is non empty, $\gamma(x)$ is contained in E, so g is defined. We first show that $r_x(z)$ is well defined: Suppose $g_1, g_2 \in \gamma(x) \cap \gamma(z)$. We have to show $[z, s(g_1)] = [z, s(g_2)]$, i.e. $s(g_1)^{-1} s(g_2) \in s(E)$ (see 3.6). But $x \in g_1F \cap g_2F \cap F$, since $g_1, g_2 \in \gamma(x)$, so $s(g_1)^{-1} s(g_2) = s(g_1^{-1} g_2) \in s(E)$ by (3.1).

3.16 (a): For $z \in F$ we have $e \in \gamma(x) \cap \gamma(z)$, so $r_x(z) = [z, s(e)] = [z, e] = t(z)$.

3.16 (b): Suppose x_1, x_2 are points of F, $h = s(g) \in s(E)$, such that $x_2 = hx_1 = gx_1$. We have to show that r_{x_2} and $s(g)^r x_1$ have the same germ at x_2. We actually show that the two maps coincide where both are defined, namely on $V = U_{x_2} \cap gU_{x_1}$. Suppose $z \in V$, $g_2 \in \gamma(x_2) \cap \gamma(z)$. Then $g^{-1}g_2 \in \gamma(x_1) \cap \gamma(g^{-1}z)$ by 4.2. So $r_{x_2}(z) = [z, s(g_2)]$, $s(g)^r x_1(z) = [z, s(g)s(g^{-1}g_2)]$. But $x_2 \in gF \cap g_2F \cap F$, so 3.1 implies $r_{x_2}(z) = s(g)^r x_1(z)$.

3.16 (c): Let $x_1 \in F$, $x \in U_{x_1}$, $h^{-1} = s(g)$, where $g \in \gamma(x_1) \cap \gamma(x)$, so $hx = g^{-1}x = x_2 \in F$. We have to show that r_{x_2} and $h^r x_1$ have the same germ at x_2 or that $s(g)^r x_2$ and r_{x_1} have the same germ at x. Let V be a neighbourhood of x such that $V \subset (\gamma(x) \cap \gamma(x_1))F$ and $V \subset U_{x_1}$, $V \subset gU_{x_2}$. We show that the two sections coincide on V: Let $z \in V$. By definition there is a $g_1 \in \gamma(x) \cap \gamma(x_1) \cap \gamma(z)$. So $r_{x_1}(z) = [z, s(g_1)]$. On the other hand $g^{-1} g_1 \in \gamma(x_2) \cap \gamma(g^{-1}x_1) \cap \gamma(g^{-1}z)$. So $_{s(g)}r_{x_2}(z) = [z, s(g)s(g^{-1}g_1)]$. But $x_1 \in gF \cap g_1F \cap F$, so $s(g^{-1}g_1) = s(g)^{-1}s(g_1)$.

4.4 LEMMA. _Either one of the following conditions implies the hypothesis of lemma_ 4.3:

(1) _Every point_ $x \in X$ _has a neighbourhood_ U_x _such that_ $\phi \neq \gamma(y) \cap \gamma(x)$ _for every_ $y \in U_x$.

(2) _The set_ $\{gF; g \in G\}$ _is a locally finite cover of_ X _and_ F _is closed_.

(3) $G\overset{\circ}{F} = X$.

PROOF. (1): We may assume that U_x is open. The hypothesis implies: $\gamma(x)F$ is a neighbourhood of x for every $x \in X$. Now for $y \in U_x$: $(\gamma(y) \cap \gamma(x))F = \gamma(y)F$ is a neighbourhood of y.

(2) is a special case of (1): Every point $x \in X$ has a neighbourhood U such that $gF \cap U$ implies $x \in gF$, which is equivalent to (1).

(3) is obvious.

Putting everything together, we obtain the following Theorem which implies Theorems 1 and 2:

4.5 THEOREM. _Suppose_ (1.2.1) _through_ (1.2.4) _hold,_ X _is connected, and any point of_ $x \in X$ _has a neighbourhood_ U_x _such that_ $(\gamma(x) \cap \gamma(y))F$ _is a neighbourhood of_ y _for every_ $y \in U_x$ (e.g. _if one of the conditions in_ 4.4 _holds_). _Then_ $q: H \to G$ _is surjective_.

There is a regular connected covering space $p:Y \to X$ *with the following properties:*

(1) H *acts on* Y *and* $p:Y \to X$ *is an* H-*map*.

(2) $K = \ker q$ *is the group of covering transformations of* p. *The group* K *acts transitively on the fibres of* p.

(3) *There is a continuous section for* p *over* F.

PROOF. The surjectivity of q was proved in 2.1. Note that the hypothesis of 4.3 implies that EF is a neighbourhood of F. Let $p:Y \to X$ be as in section III. Since p is an H-map and $GF = X$, p is surjective. Endow Y with the topology from 3.16. Then p is a covering map enjoying the properties (1) through (3) - except for the definite article in (2). We show that Y is connected: Since F is connected, so is $t(F)$, hence also $s(E)t(F)$, since $s(g)t(F) \cap t(F) \neq \emptyset$ for $g \in E$ (see 3.11). Inductively, $s(E)^n t(F)$ is connected so $Y = Ht(F) = \bigcup s(E)^n t(F)$ is connected. If Y is connected, the group of covering transformations acts freely on Y. This justifies the definite article in (2).

4.6 EXAMPLE. [10] Let T be a simplicial complex, let the group G act simplicially on T. Suppose there is a subcomplex T' of T such that for every simplex $s \in T$ there is a simplex $s' \in T'$ and a $g \in G$ with $gs' = s$. Let $X = |T|$ be the geometric realization of T, $F = |T'| \subset |T|$. If $\pi_0(F) = \pi_0(X) = \pi_1(X) = 0$, $q:H \to G$ is an isomorphism.

PROOF. We show that 4.4 (1) holds. The geometric realization $|T|$ is the set of functions λ from the set T^o of vertices of T to $[0,1]$ such that the support of λ, $\mathrm{supp}(\lambda) = \{i \in T^o;\ \lambda(i) \neq 0\}$, is a simplex of T and $\Sigma\, \lambda(i) = 1$, $(i \in T^o)$. By definition $g\lambda = \lambda o g^{-1} : T^{o} \to [0,1]$. So $g\lambda \in |T'|$ implies $g\mu \in |T'|$ for every μ with $\mathrm{supp}(\mu) \subseteq \mathrm{supp}(\lambda)$. Now $U_{\mu} = \{\lambda \in T;\ \mathrm{supp}(\mu) \subseteq \mathrm{supp}(\lambda)\}$ is an open neighbourhood of μ in the weak topology and 4.4 (1) holds for these neighbourhoods.

V PROOF OF THEOREM 3

We need two preparatory sections.

5.1 FUNDAMENTAL GROUPOID OF G-SPACES. Let X be a G-space. The group G acts on the fundamental groupoid $\pi_1(X)$ of X. We can define the semi-direct product.

$$G \ltimes \pi_1(X) =: \pi_1(G,X)$$

which is again a groupoid (cf. [3]): Its objects are $\mathrm{Ob}(\pi_1(G,X)) = \mathrm{Ob}(\pi_1(X)) = X$. For $a,b \in X$ define the set of morphisms from a to b:

$$\pi_1(G,X)(a,b) = \{(q,\alpha);\ g \in G,\ \alpha \in \pi_1(X)(a,b)\}$$

so α is a homotopy class of paths from a to b. Composition of morphisms $(g,\alpha): a \to b$ and $(h,\beta): b \to c$ is defined as

$$(h,\beta)\ (g,\alpha) = (h\ g,\ \beta + h o \alpha),$$

where + is composition of homotopy classes of paths. Obviously π_1 is a functor from G-spaces to groupoids. If f_0, f_1 are G-homotopic G-maps $X \to Y$, there is a natural equivalence of the functors $f_{0*}, f_{1*} : \pi_1(G,X) \to \pi_1(G,Y)$, here the groupoids are regarded as categories.

Let p be a point of X. Let $\pi_1(G,X;p) = \pi_1(G,X)(p,p)$ be the group of morphisms from p to p. This is the fundamental group of a transformation group considered by Rhodes [7], cf [4; §3].

Projection to the first factor in $\pi_1(G,X;p)$ yields an exact sequence

$$E(X,p) \qquad 1 \to \pi_1(X;p) \to \pi_1(G,X;p) \to G \to 1$$

if X is pathconnected. A G-map $f: X \to Y$ induces a map $E(X,p) \to E(Y,f(p))$ of the resp. extensions of G, for any $p \in X$. If f_0, f_1 are G-homotopic G-maps $X \to Y$, there is a horizontal isomorphism of extensions in the commutative diagram

$$\begin{array}{ccc} & E(X,p) & \\ {}^{\pi_1(f_0)}\swarrow & & \searrow^{\pi_1(f_1)} \\ E(Y,f_0(p)) & \cong & E(Y, f_1(p)). \end{array}$$

5.2 NERVES OF COVERS. This is almost literally a part of [8], for a more detailed version see [2; pp.224-227]. Let X be a topological space, $U = \{U_\alpha\}_{\alpha \in \Sigma}$ be a <u>cover</u> of X by subsets U_α. If σ is a subset of Σ define $U_\sigma = \bigcap_{\alpha \in \sigma} U_\alpha$. Let R_U be the category whose objects are the nonempty U_σ for finite subsets σ of Σ and whose morphisms are their inclusions. The nerve NR_U of the category R_U is the barycentric subdivision of what is ordinarily called the <u>nerve</u> of U.

There is also another category X_U associated to U. It is a topological category whose objects are the pairs (x, U_σ) with $x \in U_\sigma$, and whose morphisms $(x, U_\sigma) \to (y, U_\tau)$ are inclusions $i : U_\sigma \hookrightarrow U_\tau$ such that $i(x) = y$. More formally $ob(X_U) = \coprod_\sigma U_\sigma$, the sum being taken over all finite subsets of Σ, and $mor\ (X_U) = \coprod_{\sigma \subset \tau} U_\tau$, with the sum over all pairs of finite subsets $\sigma \subset \tau$ of Σ.

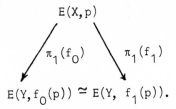

The geometric realization of the nerve of a topological category C is called the classifying space BC of the category C.

Returning to the notations of 1.1, let U be the cover $\{gF;\ g \in G\}$ of X. BR_U and BX_U are G-spaces. We abbreviate

$$\pi_1(G;F) := \pi_1(G, BR_U;\ e),$$

$e \in BR_U$ the neutral element of G.

Let H be the group with generators s(g), $g \in E$, and relations 1.2.1.

5.3 PROPOSITION. <u>There is a vertical isomorphism in the commutative diagram</u>

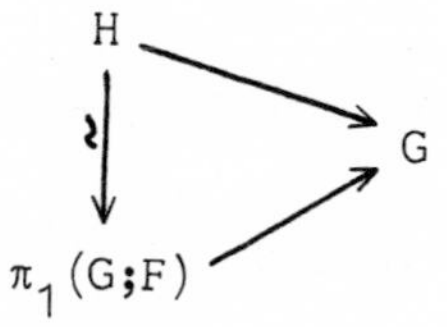

PROOF. $g \in E$ if and only if $(e,g) \in R_U$. So in BR_U there is a line from g to (e,g) and one from (e,g) to e. Let us call $\sigma(g)$ the homotopy class of the composite path from $g \in E$ to e.

Let F(E) be the free group with free base E. We have a commutative diagram of group homomorphisms

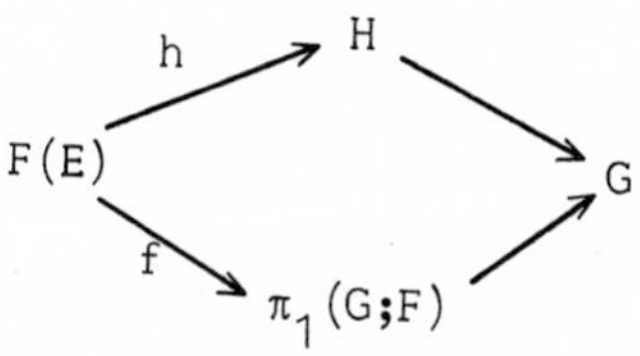

where f is defined by

$$f(g) = (g, \sigma(g)) \qquad \text{for } g \in E$$

and h is defined by

$$h(g) = s(g) \qquad \text{for } g \in E.$$

Now it is an easy consequence of the edge-path description of $\pi_1(G,BR_U;e) = \pi_1(G;F)$ and the hypotheses and definitions that there is a vertical isomorphism making the completed diagram commutative.

5.4 THEOREM. *Let* $\{gF; g \in G\}$ *be a G-numerable cover of the G-space* X. *Suppose* H *is the group with set of generators* $\{s(g); g \in E\}$ *and relations* 1.2.1. *If* $\pi_o(F) = \pi_o(X) = 0$ *we have a map of exact sequences*

$$\begin{array}{lccccccccc} & 1 & \to & \pi_1(X,*) & \to & \pi_1(G,X;*) & \to & G & \to & 1 \\ & & & \downarrow{\scriptstyle \nu} & & \downarrow & & \downarrow{\scriptstyle id} & & \\ \text{(E)} & 1 & \to & K & \to & H & \xrightarrow{q} & G & \to & 1. \end{array}$$

Here $*$ *is a basepoint* $\in F$. *The vertical arrows are surjective. The kernel of* ν *is a normal subgroup of* $\pi_1(X,*)$ *containing the image of* $\pi_1(F,*) \to \pi_1(X,*)$ *and is G-stable in the following sense: If* $[c] \in \ker(\nu)$, $g \in G$ *and* d *is a path from* $g*$ *to* $*$, *then the homotopy class of the composite path* $d^- + goc + d$ *is in* $\ker \nu$.

This theorem obviously implies Theorem 3.

PROOF. The obvious functor $X_U \to R_U$ induces a G-map $BX_U \to BR_U$. If F is pathconnected, so is BX_U. So we have a map of exact sequences

$$E(BX_U, (e,*)) \to E(BR_U, e)$$

which is surjective, since $\pi_1(G,BR_U;e) = \pi_1(G;F)$ is generated by $(g,\sigma(g))$ and $\pi_o(F) = 0$. Proposition 5.3 gives an isomorphism $E \cong E(BR_U, e)$. The G-map $BX_U, (e,*) \to X, *$ is a G-homotopy equivalence if U is a G-numerable cover (cf. [8; 4.1] or [2; pp.224-227], so induces an isomorphism $E(BX_U, (e,*)) \xrightarrow{\cong} E(X,*)$. So the missing arrow in the diagram

$$\begin{array}{ccc} E(X,*) & \xleftarrow{\simeq} & E(BX_U,(e,*)) \\ & & \downarrow \\ E & \xleftarrow{\sim} & E(BR_U, e) \end{array}$$

yields the map of the Theorem. Everything else follows now.

REFERENCES

1. H. Behr. Über die endliche Definierbarkeit von Gruppen. _Journal f. reine u. ang. Math._ 211 (1962), 116-122.
2. J. M. Boardman and R. M. Vogt. _Homotopy everything H spaces._ Lecture Notes in Maths. 347, Springer, Berlin-Heidelberg-New York 1973.
3. R. Brown. Groupoids as coefficients. _Proc. London Math. Soc._ III 25 (1972), 413-426.
4. R. Brown and G. Danesh-Naruie. The fundamental groupoid as a topological groupoid. _Proc. Edinburgh Math. Soc._ 19 (1975), 237-244.
5. M. Gerstenhaber. On the algebraic structure of discontinuous groups. _Proc. Amer. Math. Soc._ 4 (1953), 745-750.
6. A. M. Macbeath. Groups of homeomorphisms of a simply connected space. _Ann. of Math._ 79 (1964), 473-488.
7. F. Rhodes. On lifting transformation groups. _Proc. Amer. Math. Soc._ 19 (1968), 905-908.
8. G. Segal. Classifying spaces and spectral sequences. _Publ. Math. IHES_ 34 (1968), 105-112.
9. C. L. Siegel. Discontinuous groups. _Ann. of Math._ 44 (1943), 674-689 = Ges.Abh.II 390-405.

10. C. Soulé. Groupes opérant sur un complexe simplicial avec domaine fondamental. C.R. Acad. Sc. Paris 276 (1973) A 607-609.

11. R. G. Swan. Generators and relations for certain special linear groups. Advances in Math. 6 (1971), 1-77.

UNIVERSITY OF BIELEFELD,
BIELEFELD, WEST GERMANY.

AFFINE EMBEDDINGS OF REAL LIE GROUPS

NGUIFFO B. BOYOM

INTRODUCTION

Let G be a connected real Lie group. This paper is the first part of a work devoted to the study of 'maximal' orbit types of smooth G-actions in smooth manifolds equipped with a fixed linear connection (so called affine smooth manifold). In this paper, let us limit ourselves to the special case of the real flat euclidian manifold R^p.

The first and second sections are concerned with preliminaries and examples. In particular the examples illustrate the natural problems.

A semi-affine orbit is an orbit contained in a proper affine submanifold. In section 3, the semi-affine orbit existence problem is solved by means of the notion of vanishing cohomology classes.

In section 4, the affine embedding theorem is proved.

In the fifth and last section I try to outline an algebraic interpretation of some special linear orbits of affine embeddings.

1. SOME PRELIMINARIES

All manifolds we consider are smooth, paracompact and

connected. Let M be a manifold and let us consider a connected real Lie group G which acts smoothly (on the left) in the manifold M, that is one is given a smooth map φ from the product manifold $M \times G$ into the manifold M with the two following properties: let (s,s') be a pair of elements of G and let m be any point in M, then

1) $\varphi(s,\varphi(s',m)) = \varphi(ss',m)$

2) $\varphi(e,m) = m$

where e is the unit element of the group G. The stability subgroup of a point $m \in M$ is denoted G_m and its orbit is denoted $G(m)$.

Suppose, for example, that φ is a locally free action. Let D_φ be the foliation of M generated by the one parameter subgroups in G. The leaves of D_φ are nothing but the orbits of φ, thus the orbit map φ_m from G into $G(m)$ defined by

$$\varphi_m(s) = \varphi(s,m)$$

is continuous. This fact implies the pair (G,φ_m) is a smooth covering of the manifold $G(m)$.

Now suppose the manifold $G(m)$ is a flat one, then any locally flat linear connection on $G(m)$ may be lifted to a locally flat connection on G. If the action $\varphi_{G(m)}$ is affine flat action, that means there is a G-invariant locally flat connection on $G(m)$, the locally flat connection lifted on G defines a left symmetric algebra structure on the Lie algebra $\mathfrak{g}$ of the Lie group G, [3]. So sufficient conditions for the hyperbolicity of the orbit $G(m)$ may be obtained from the Koszul-Vinberg algebraic models for the theory of homogeneous convex domains,[1,4]. The relation between such global differential geometric objects on Lie groups (invariant linear connections) and the theory of affine representations of Lie algebras are well known [3]. In general, if M is a smooth G-manifold we always suppose the dimension of M is at least equal to that of G.

Let G be a connected real Lie group. Suppose M is a smooth manifold equipped with a smooth G-action φ. If there is at least one point m_o in M the stability subgroup of which is discrete then the action φ is said <u>to realize an immersion</u> of G into M. The orbit $G(m_o)$ is called a <u>maximal orbit</u>. We always suppose the orbit map φ_{m_o} is smooth. In what follows we are concerned with G-action in real euclidian manifolds R^p.

2. AFFINE ACTIONS OF REAL LIE GROUPS IN EUCLIDIAN MANIFOLD

Let G be a connected real Lie group, consider M to be the euclidian real space equipped with its natural smooth structure. There is a unique linear connection ∇ defined on M such that the covariant derivative of every constant vector field vanishes.

An affine action of the group G in M, or affine representation, is a pair (q,f) where:

(a) f is a continuous linear representation of the Lie group G on the vector space M, and

(b) q is a smooth f-cocycle of dimension one.

The last property means that q is a smooth map from G into M which satisfies the following: given a pair (s,s′) of elements of G and a point m in M one has

3) $q(ss') - q(s) = f(s)\, q(s')$.

If (q,f) is an affine action of the real Lie group G on M every point m in M has a (q,f)-orbit denoted by

$$G_a(m) = \{q(s) + f(s)\, m, s \in G\}$$

and an f-orbit denoted by

$$G_\ell(m) = \{f(s)\, m, s \in G\} .$$

In this paper we deal with some (q,f)-orbits of a special nature which will be made clear in the following sections. Suppose $\dim G < \dim M$, it is not difficult to see that not every affine action (q,f) of G in M has a (q,f)-orbit contained in a proper affine submanifold. This problem will be discussed.

We shall give a necessary and sufficient condition for this fact to occur and exhibit some properties of the affine actions (q,f) which realize immersions and a maximal (q,f)-orbit which is a proper affine submanifold. Before doing that we need the

DEFINITION 1.1. One says the affine action (q,f) of G in M realizes an embedding of G into M if the maps q and f are immersions in the usual sense.

Some of the problems solved in this paper may be illustrated by the following three examples.

EXAMPLE 1. Let G be the simply connected 3-dimensional real Lie group the elements of which are real 2 x 2 matrices $S = \begin{bmatrix} \lambda & \nu \\ 0 & \mu \end{bmatrix}$, where λ and μ are elements of the multiplicative group R^+ and $\nu \in R$. Consider the affine action of G on $M = R^3$ defined by the pair

$$q(s) = \begin{bmatrix} \nu \\ \mu-1-\log \lambda \\ \log \lambda \end{bmatrix} \qquad f(s) = \begin{bmatrix} \lambda & \nu & \nu \\ 0 & \mu & \mu^{-1} \\ 0 & 0 & 1 \end{bmatrix}$$

(Every euclidean space is equipped with its canonical basis.) The pair (q,f) realizes an embedding of the group G in M according to the Definition 1.1.

One observes this action is not transitive on M. It is trivial (but crucial) to see that any affine action (q,f) of R^+ in the one-dimensional manifold R that realizes an embedding of R^+ in R is given by

$$q(\lambda) = \alpha(\lambda-1), \quad \alpha \in R,$$
$$f(\lambda) = \lambda$$

and is never transitive on R. This remark shows that not every affine action (q,f) realizing an embedding has an 'affine orbit'.

EXAMPLE 2. Take G to be the real one-dimensional torus T^1, and make it act on the manifold $M = R^2$ by rotation around the origin of R^2. It is easy to show that the one-dimensional cohomology space $H^1(T^1,R^2)$ is trivial. In fact given two elements $e^{i\theta}$ and $e^{i\theta'}$ in T^1 and a one-dimensional cocycle q, the relation (3) gives

$$3')\qquad e^{i\theta}\, q(e^{i\theta'}) - q(e^{i\theta'}) = e^{i\theta'}\, q(e^{i\theta}) - q(e^{i\theta}) .$$

By partial differentiation with respect to θ' we obtain

$$q(e^{i\theta}) = \delta[d\, q(e^{i3\pi/2})]\ (e^{i\theta})$$

where δ is the coboundary operator; thus $H^1(T^1,R^2) = 0$. The vanishing of $H^1(T^1,R^2)$ implies that all the orbits of the affine T^1-action in R^2 are circles. So no such orbit is contained in a proper affine submanifold.

EXAMPLE 3. Let G be the multiplicative group R^+ and take the affine G-action in R^2 given by the pair

$$q(\lambda) = \begin{bmatrix} \log \lambda \\ \lambda - 1 \end{bmatrix} \quad , \quad f(\lambda) = \begin{bmatrix} 1 & 0 \\ 0 & \lambda \end{bmatrix} .$$

Because of irrationality of the function log, the cohomology space $H^1(G,R^2)$ is not trivial. In particular the cohomology class of the cocycle q above does not vanish.

Now let $m = \begin{bmatrix} x_1 \\ x_2 \end{bmatrix}$ be a point in R^2, the orbit of m is the set

$$G_a(m) = \left\{ \begin{bmatrix} x_1 + \log \lambda \\ \lambda(x_2+1) - 1 \end{bmatrix} \quad , \lambda \in R^+ \right\} .$$

One sees the orbit of any $m = \begin{bmatrix} x_1 \\ -1 \end{bmatrix}$ is the affine submanifold

$$\left\{ \begin{bmatrix} \alpha \\ -1 \end{bmatrix} \quad , \alpha \in R \right\} .$$

Moreover the pair (q,f) realizes an embedding of G in R^2, however, the restriction of (q,f) on the orbit $G\begin{bmatrix} x \\ -1 \end{bmatrix}$ does not realize an embedding.

In the next section it will be clear that for affine G-action the existence of a (q,f)-orbit contained in a proper affine submanifold has a topological meaning.

3. THE VANISHING THEOREM FOR COHOMOLOGY CLASSES

Given a connected real Lie group G, let V and W be two G-modules of finite dimension. Any morphism α from the G-module V to the G-module W induces an R-linear homomorphism α^* of degree zero from the cohomology space $H^*(G,V)$ to the cohomology space $H^*(G,W)$.

In this section the morphisms of G-modules we consider are epimorphisms. The notion of vanishing cohomology class gives us an effective criterium for deciding whether or not an affine action (q,f) possesses an orbit contained in proper affine submanifold. Let us set

DEFINITION 3.1. Let V be a G-module of finite dimension, a cohomology class $[q] \in H^t(G,V)$ is a vanishing class if there exists a non-null G-module W and an epimorphism α from V to W such that $[q] \in \operatorname{Ker} \alpha^*$.

REMARK. Consider monomorphism instead of epimorphisms, then one gets the usual Iwasawa-Hochschild-Koszul vanishing class theory (effacement des classes des cohomologie). However, the geometrical interpretation we obtain shows that not every cohomology class in $H^1(G,V)$ is a vanishing class in the above sense although it is with respect to the I.H.K. vanishing theory. In particular if G is a solvable Lie group, any $[q] \in H^t(G,V)$ is an I.H.K. vanishing class.

Now let (q,f) be an affine action of a connected real Lie group G in $M = R^p$, with $p > \dim G = r$. Suppose that the pair (q,f) realizes an immersion of G in M.

DEFINITION 3.2. We say the pair (q,f) realizes an affine immersion of G in M if there is a maximal orbit $G_a(m_o)$ that is proper affine submanifold.

With respect to the theory of homogeneous convex domains such an affine orbit is the most 'degenerate' one.

The main result of this section is the following.

VANISHING THEOREM (3.1) <u>Let</u> (q,f) <u>be an affine action of a connected real Lie group</u> G <u>in</u> R^p, $p > \dim G = r$; <u>then there is an orbit contained in an r-dimensional affine submanifold if and only if the</u> [q] <u>is a vanishing class</u>.

PROOF. The condition is necessary: in fact let m_o be a point in M the orbit of which is contained in the r-dimensional affine submanifold $m_o + V$, V being a subspace of the vector space $M = R^p$. Any point m of the orbit $G_a(m_o)$ splits into

$$m = m_o + m'$$

where m′ is an element of the vector space V. So take an element S in G, then we get

$$\begin{aligned} 4) \qquad s\,m &= q(s) + f(s)\,m \\ &= q(s) + f(s)\,m_o + f(s)\,m' \\ &= s\,m_o + f(s)\,m' \end{aligned}$$

Since s m lies in $m_o + V$ for the point m in $G_a(m_o)$ one obtains

$$4') \qquad s\,m - s\,m_o = f(s)\,m'$$

which clearly lies in the vector space V. Now, since $G_a(m_o)$ is open in $m_o + V$, the relation (4′) implies that V is a sub-G-module of M. Consider the factor space W = M/V and let us denote by $\tilde{f}$ the continuous linear representation of G in W induced by f; α being the canonical epimorphism from the G-module M into W, the $\tilde{f}$-cocycle $\alpha \circ q$ is cohomologous to zero. This is a direct consequence of (4), since

$$\begin{aligned} \alpha\, q\,(s) &= \alpha\, m_o - \alpha\, f(s)\, m_o \\ &= \alpha\, m_o - \tilde{f}(s)\,\alpha\, m_o \\ &= \delta\,[-\alpha\, m_o]\,(s) \end{aligned}$$

The condition is sufficient: suppose there is a G-module W and an epimorphism α from M into W such that $\alpha^*[q] = 0$. The affine representation $(\alpha \circ q, f')$ of G in W (f′ being the linear representation which defines the G-module structure on W) is conjugated by an affine automorphism of W to its linear part f′. This assertion is equivalent to the existence of a fixed point w_o (with respect to $(\alpha \circ q, f'))$. Since α is a morphism of G-modules the affine submanifold $\alpha^{-1}(w)$ is (q,f)-stable. Let V be the homogeneous part of $\alpha^{-1}(w_o)$, V is the kernel of α, so is a submodule of M. Now it is obvious that the (q,f)-orbit of any point in $\alpha^{-1}(w_o)$ lies in $\alpha^{-1}(w_o)$.

4. AFFINE IMMERSIONS AND AFFINE EMBEDDINGS OF REAL LIE GROUPS

The main purpose of this section is to prove the affine embedding theorem for connected real Lie groups. Let us fix some conventions. Given a connected r-dimensional real Lie group G, and the Euclidian manifold $M = R^p$, $p > r$, let (q,f) be an affine G-action in M. Briefly speaking we deal with triples $(M,G,(q,f))$. Such an object will designate an affine action of G in M that may or may not realize an affine immersion. We saw that if $(M,R^+,(q,f))$ is an affine embedding of R^+ in M, then $\dim M > 1$.

Suppose we are given two triples $(M,G,(q,f))$ and $(M',G',(q',f')$

DEFINITION 4.1. A morphism from $(M,G(q,f))$ into $(M',G',(q',f'))$ is a pair (α,h) where α is an R-linear homomorphism from M into M′ and h is a continuous homomorphism from the Lie group G into G′ such that the following holds:

$P_1)$ $\alpha \circ f(s) = f'(h(s)) \circ \alpha$, $s \in G$;

$P_2)$ The two cocycles $\alpha \circ q$ and $q' \circ h$ are cohomologous.

We may remark that modulo an affine conjugation the second property may be replaced by the equality of the two cocycles $\alpha \circ q$ and $q' \circ h$.

Now let (α,h) be a morphism of triples as above, set V to be the kernel of α and H to be that of h. Obviously the subgroup H acts by translations in M'. On the other hand V is a submodule of the G-module M. Using proper (P_2) one easily proves that $q(H)$ lies in V. So, let us denote by $(q,f)_H$ the restriction of (q,f) to H. The pair (i_V,i_H) of inclusion maps defines a monomorphism from the triple $(V,H,(q,f)_H)$ into $(M,G,(q,f))$. The composed pair $(\alpha \circ i_V, h \circ i_H)$ from $(V,H,(q,f)_H)$ into $(M',G',(q',f'))$ is a null one with the obvious meaning.

Thus, one gets a natural notion of sequence and exact sequence of triples: $\cdots \longrightarrow (M_r,G_r,(q_r,f_r)) \xrightarrow{(\alpha_{r+1},h_{r+1})} (M_{r+1},G_{r+1},(q_{r+1},f_{r+1})) \longrightarrow \cdots$ Now given a short exact sequence of triples:

$$5) \qquad (M_o,G_o,(q_o,f_o)) \xrightarrow{(\alpha,h)} (M,G,(q,f)) \xrightarrow{(\alpha_1,h_1)} (M_1,G_1,(q_1,f_1)),$$

(that means the pairs (α,h) and (α_1,h_1) are respectively a monomorphism and epimorphism), we say that (5) is an <u>extension</u> of $(M_1,G_1,(q_1,f_1))$ by $(M_o,G_o,(q_o,f_o))$. Let us say the triple $(M,G,(q,f))$ is <u>trivial</u> if f is the trivial linear representation.

Before giving the proof of the embedding theorem, let us start with the following elementary result:

LEMMA 4.0. <u>Let</u> $(M,G,(q,f))$ <u>be a triple that realizes an affine immersion, then we can make the orbit of the origin</u> $G_a(0)$ <u>a maximal affine one</u>.

The proof is easy, so we omit it.

It follows from the above Lemma that the pair $(G,(q,f)_o)$ is a covering manifold of the affine submanifold $G_a(0)$, then the

Lie group G is simply connected. Finally we see that if $(M,G,(q,f))$ realizes an affine immersion then the triple $(V = G_a(0),G,(q,f))$ also does.

AFFINE EMBEDDING THEOREM 4.1. _Let_ $(M,G,(q,f))$ _be an affine immersion of a connected real Lie group_ G _in_ $M = R^p$. _Then the affine immersion_ $(G_a(0),G,(q,f))$ _is extension of an affine embedding_ $(\tilde{M},\tilde{G},(\tilde{q},\tilde{f}))$; _otherwise_ G _is solvable_.

COMMENT. The above theorem means that if the pair (q,f) realizes an affine immersion of G, one can find an affine embedding of a factor group of G and an epimorphism (α,h) from $(G_a(0),G,(q,f))$ into that embedding. If for some affine immersion of G such an affine embedding does not exist then the group G is a solvable Lie group.

PROOF OF THE AFFINE EMBEDDING THEOREM. Let $(M,G,(q,f))$ be an affine immersion of G, set $V = G_a(0)$ and f_V to be the continuous linear representation of G in V induced by f. The pair (q,f_V) realizes an affine immersion of G in V. Now set H_1 to be the connected component of the unit element of the subgroup Ker (f_V) (not Ker (f)!).

LEMMA 4.1. (a_1) H_1 _is a simply connected abelian Lie subgroup_;

(a_2) $q(H_1)$ _is a sub-module of the G-module_ V.

PROOF. From (3) one gets:

6) $q(ss') = q(s) + q(s')$

for every pair (s,s') of elements in H_1. Let a_t be a one parameter subgroup in H_1, we easily see that for any rational number r, one gets:

$$q(a_{rt}) = r\, q(a_t) \ .$$

Making use of the continuity argument we obtain:

$$q(a_t) = t\, q(a_1) \ .$$

So, $q(H_1)$ is R-linearly generated by some $q(a_1^{\,k})$, $k = 1,\ldots,\dim H_1$, where the a_t^k are one parameter subgroups in H_1. That proves (a_1).

Now let S be an element in G, and $\sigma \in H_1$, we have:

$$q(s\,\sigma\, s^{-1}) = f_V(s)\, q(\sigma) \ ,$$

this means the G_{f_V}-stability of $q(H_1)$, which proves (a_2).

That being, let us denote by M_1 the factor module $V/q(H_1)$ and $\tilde{f}_1$ the factor linear representation of G in M_1 induced by f_V. Set G_1 the factor group $H_1 \backslash G$. Let α_1 and h_1 be the canonical epimorphisms of V into M_1 and of G into G_1 respectively. It is clear that $\tilde{f}_1$ gives rise to a linear (continuous) representation f_1 of G_1 in M_1 such that:

$$\tilde{f}_1(s) = f_1(h_1(s)) \ , \quad s \in G \ .$$

There exists an f_1-cocycle q_1 making the following diagram commutative.

7)
$$\begin{array}{ccc} G & \xrightarrow{\ q\ } & V \\ {\scriptstyle h_1}\downarrow & & \downarrow{\scriptstyle \alpha_1} \\ G_1 & \xrightarrow{\ q_1\ } & M_1 \end{array}$$

Now it is easy to see that the pair (q_1, f_1) realizes an affine immersion of G_1 in M_1. Moreover the pair (α_1, h_1) is an epimorphism of triples.

By iterating the process, this leads to a family of epimorphisms:

8) $$(M_r, G_r, (q_r, f_r)) \xrightarrow{(\alpha_{r+1}, h_{r+1})} (M_{r+1}, G_{r+1}, (q_{r+1}, f_{r+1}))$$

and (8) defines a projective system of triples. Then set

$$(\widetilde{M}, \widetilde{G}, (\widetilde{q}, \widetilde{f})) = \varprojlim (M_r, G_r, (q_r, f_r)) .$$

On the other hand (7) gives rise to a normal series

9) $$e = H_0 \subset H_1 \subset H_2 \subset \dots \subset H_r \subset \dots$$

such that H_r contains the commutator subgroup of H_{r+1}. Then the limit $H = \varinjlim H_r$ is a solvable subgroup of G. The following two assertions are equivalent:

(A_1) $H \neq G$,

(A_2) $\dim \widetilde{G} > 0$ and $(\widetilde{M}, \widetilde{G}, (\widetilde{q}, \widetilde{f}))$ is an affine embedding.

In fact there is no difficulty to see that $\widetilde{G} = H \backslash G$. So if one of ($A_1$), ($A_2$) does not hold then G is a solvable group.

The following is an immediate consequence of the above Theorem.

COROLLARY. _Let_ G _be a connected real non-solvable Lie group. For any affine immersion_ (q, f) _of_ G, _there exists a connected real Lie group of strictly positive dimension_ $\widetilde{G}$, _an affine embedding_ $(\widetilde{M}, \widetilde{G}, (\widetilde{q}, \widetilde{f}))$ _and an epimorphism_ (α, h) _from_ $(G_a(0), G, (q, f))$ _into_ $(\widetilde{M}, \widetilde{G}, (\widetilde{q}, \widetilde{f}))$.

Both conclusions of the above theorem may hold simultaneously. This is illustrated by a simple (but non-trivial) example.

EXAMPLE 4. Take M to be the 3-dimensional real euclidian space, equipped with its canonical basis e_1, e_2, e_3. Now let G be the connected 3-dimensional real Lie group the elements of which are 3 x 3 matrices:

$$S = \begin{bmatrix} 1 & \lambda & \mu \\ 0 & 1 & \nu \\ 0 & 0 & 1 \end{bmatrix} ,$$

where $(\lambda,\mu,\nu) \in R^3$. Let $\mathfrak{g}$ be the Lie algebra of G. The elements of $\mathfrak{g}$ are the 3 x 3 real matrices:

$$X = \begin{bmatrix} 0 & \lambda & \mu \\ 0 & 0 & \nu \\ 0 & 0 & 0 \end{bmatrix} .$$

Let us denote by $\overline{f}$ the linear representation of $\mathfrak{g}$ in R^3 defined by:

$$\overline{f}(X) = \begin{bmatrix} 0 & -\lambda & -\mu \\ 0 & 0 & \nu \\ 0 & 0 & 0 \end{bmatrix} , \quad X \in \mathfrak{g} .$$

Direct computation shows that the 1-dimensional cohomology space $H^1(\mathfrak{g},R^3)$ is not zero. In particular take the cocycle $\overline{q}$ defined by:

$$\overline{q}(X) = (\mu + \nu)e_1 + (\mu - \nu)e_2 + \lambda e_3 ,$$

then $\overline{q}$ is not cohomologous to zero.

Let (q,f) be the unique affine G-action in M the differentia of which at the unit element is the pair $(\overline{q},\overline{f})$.

PROPOSITION 4.1. _The triple_ $(M,G,(q,f))$ _is an affine embedding of_ G _in_ R^3.

PROOF. It is clear the pair (q,f) realizes an embedding of G. It remains to prove this is an affine embedding. This fact is obtained by proving the transitivity of (q,f). We claim that any point m in M has open orbit. To prove this it suffices to verify it locally. In fact choose arbitrarily a point $m_o = xe_1 + ye_2 + ze_3$ in M and take the differential at the unit element of the orbit map $(q,f)_{m_o}$, you get the pair $(\overline{q},\overline{f})_{m_o}$. So the image of the element

$$X = \begin{bmatrix} 0 & \lambda & \mu \\ 0 & 0 & \nu \\ 0 & 0 & 0 \end{bmatrix}$$

is the point $(\lambda y-\mu z+\mu+\nu)e_1 + (\nu z+\mu-1)e_2 + \lambda e_3$. Then the matrix of $(\overline{q},\overline{f})_{m_o}$ is:

$$\begin{bmatrix} -y & 1-z & 1 \\ 0 & 1 & z-1 \\ 1 & 0 & 0 \end{bmatrix} .$$

This matrix is always regular and the claim is proved.

5. KOSZUL-VINBERG ALGEBRAS WITH RIGHT UNITS

This section is concerned with some elementary properties of linear orbits $G_\ell(\ )$ of an affine embedding. We make use of results that can be found in the literature, so proofs will be ommited.

LEFT SYMMETRIC ALGEBRAS. Let k be a commutative field. Let us recall that a left symmetric algebra (Koszul-Vinberg algebra) is a k-algebra V the associator of which is left symmetric. That is to say, for every triple (X,Y,Z) of elements in V one gets

$$(XY)Z - X(YZ) = (YX)Z - Y(XZ) .$$

It is well-known that the commutator algebra of a left symmetric algebra is a Lie algebra.

Now let k be the field of real numbers. A left symmetric algebra is of <u>compact type</u> if there exists on it a positive definite quadratic form. A <u>clan</u> is a left symmetric algebra of compact type such that all the left multiplication operators have only real eigenvalues. These algebra structures play a crucial role in the theory of convex homogeneous domains, (domains considered do not contain straight lines) [1, 4]. It is known that if Ω is a convex homogeneous domain in affine real space R^p, then the Lie algebra $\mathfrak{g}(\Omega)$ of infinitesimal affine derivations of Ω is a clan. Moreover if Ω is a convex cone, then $\mathfrak{g}(\Omega)$ is a clan with unit and conversely [4]. The proof of what follows can be found in [4].

THEOREM 1 (Vinberg) There is one-to-one map of the set of convex homogeneous domains on the set of clans.

THEOREM 2 (Vinberg). Let Ω be a convex homogeneous domain. Let G be an affine subgroup that acts simply transitively in Ω, then G is solvable.

LINEAR ORBITS OF AFFINE EMBEDDINGS. Let G be a connected real Lie group and suppose (q, f) is an affine G-action in R^p. Set $(\overline{q}, \overline{f})$ to be the differential at the unit element of (q, f). The pair $(\overline{q}, \overline{f})$ is an affine representation of the Lie algebra $\mathfrak{g}$ of G. Now suppose that (q, f) is an affine immersion, then take the differential at the unit element of the pair (q, f_V) where $V = G_a(0)$ is a maximal affine orbit. It is easy to see that the cohomology class $[\overline{q}] \in H^1(\mathfrak{g}, V)$ is not zero. Moreover $\overline{q}$ is a linear isomorphism.

Thus one can define on $\mathfrak{g}$ an algebra structure by setting

10) $$X\,Y = \overline{q}^{-1}\overline{f}_V(X)\overline{q}(Y) .$$

We obtain a left symmetric algebra. The commutator algebra of (10) is the Lie algebra $\mathfrak{g}$. This means that the bracket operation in $\mathfrak{g}$ is given by

$$[X, Y] = X\,Y - Y\,X .$$

Since $[\overline{q}]$ is a non-zero class, (10) has no unit element.

The following result is a direct consequence of affine embedding theorem:

PROPOSITION 5.1. *Suppose the pair* (q,f) *realizes an affine embedding of* G, *then:*

1) *All left multiplication operators of* (10) *are singular;*

2) *No right multiplication operator of* (10) *has non-zero real eigenvalue.*

Now let us denote by $\tilde{\Omega}$ the set of elements Y in $\mathfrak{g}$ such that the right multiplication operator R_Y (of (10)) is regular. The subset $\tilde{\Omega}$ is a cone and does not contain straight lines through the origin. Set $\Omega = \bar{q}(\tilde{\Omega}) = (dg)_e(\tilde{\Omega})$. The last results we shall prove are:

PROPOSITION 5.2. *Every point* m *in* Ω *gives rise to a left symmetric algebra with right unit whose commutator algebra is* $\mathfrak{g}$.

THEOREM 5.1. *Suppose that* $Ad(G)$ *has no negative eigenvalue, then* Ω *contains the convex hull of linear orbits* $G_\ell(m)$ *of its points.*

PROOF OF PROPOSITION 5.2. Take any point m in Ω, then the map Ψ_m of $\mathfrak{g}$ in $G_a(0)$ defined by $\Psi_m(X) = \bar{f}_V(X)m$ is one-to-one linear map. This Ψ_m is the coboundary of the 0-cochain m. Thus take the multiplication in the vector space $\mathfrak{g}$ given by

$$11) \qquad XY = \Psi_m^{-1}(\bar{f}_V(X)\bar{f}_V(Y)m).$$

It defines a left symmetric algebra. Set X_o to be the unique element in $\mathfrak{g}$ such that $\Psi_m(X_o) = m$, X_o is right unit for (11).

It is obvious that if the Lie group G is nilpotent then Ω is the empty set.

Before proving Theorem 5.1, we prove

LEMMA 5.0. *Ω is stable with respect to the linear operators $f(s)$, $s \in G$.*

PROOF. In fact let m_o be a point in Ω and take its image $m = f(s)m_o$. One must show the map Ψ_m is one-to-one map of $\mathfrak{g}$ in $G_a(0)$, which follows directly from the fact that $\bar{f}(X)\ f(s) = \bar{f}(Ad(s^{-1})X)$, so the Lemma is proved.

PROOF OF THEOREM 5.1. Now take m_o and $m = f(s)m_o$ and let $u = \alpha m_o + \beta m$ be a convex combination of m_o and m. This means α and β are non-negative real numbers such that $\alpha + \beta = 1$. Look at the map Ψ_u:

$$X \longrightarrow \alpha\bar{f}(X)m_o + \beta\bar{f}(X)m\ .$$

According to the Lemma 6.0 one has

$$\Psi_u(X) = \Psi_{m_o}(\alpha X + \beta Ad(s^{-1})X)\ .$$

Since $Ad(s^{-1})$ has no negative eigenvalue the map Ψ_u is one-to-one linear map.

REFERENCES

1. J.L. Koszul. Domaines bornés et orbites des groupes de transformations affines - <u>Bull. Soc. Math. France</u>. 89 (1961), 515-533.

2. J.L. Koszul. Deformations des connexions localement plates <u>Ann. Inst. Fourier</u> 18 (1968) 103-114.

3. N.B. Boyom. Algebres symétriques à gauche et Algèbres de Li réductives, Thèse Spec. (Grenoble) 1968.

4. E.B. Vinberg. The theory of convex homogeneous cones. Trudy Moskov Mat. Obsc. 12 (1963) 303-358.

U.S.T.L.

MONTPELLIER, FRANCE.

EQUIVARIANT DIFFERENTIAL OPERATORS OF A LIE GROUP*

ROBERT DELVER

1. INTRODUCTION

Let M and N be smooth manifolds and G a Lie group acting on M. The action of G lifts naturally to an action on the jet bundle $J^k(M,N)$, $C^\infty(M,R)$ and other function spaces on M. $\mathcal{D}_k$ denotes the set of smooth differential operators of $C^\infty(M,N)$ into $C^\infty(M,R)$. $\mathcal{F} \in \mathcal{D}_k$ is called G-equivariant if $\mathcal{F}(g.u)=g.(\mathcal{F}u)$, $\forall g \in G$, $\forall u \in C^\infty(M,N)$. This notion is of importance in the theory of differential equations because if $\mathcal{F}$ is G-equivariant and $a \in C^\infty(M,R)$ is G-invariant then the action of G on $C^\infty(M,N)$ leaves the solution set of $\mathcal{F} f=a$ invariant. Let $\mathcal{D}_k^G$ be the set of G-equivariant elements of $\mathcal{D}_k$. The main result is a finiteness theorem for $\mathcal{D}_k^G$: (see also the addendum).

1.1 THEOREM. <u>If</u> G <u>is a compact Lie group acting smoothly on a smooth compact manifold</u> M <u>and</u> N <u>is a smooth manifold which is countable at infinity then</u>, <u>for all</u> $k \geq 1$, <u>there exist</u> $\mathcal{B}_1,\ldots,\mathcal{B}_i \in \mathcal{D}_k^G$ <u>such that</u> $\mathcal{F} \in \mathcal{D}_k^G$ <u>if and only if</u> $\mathcal{F} = f(\mathcal{B}_1,\ldots,\mathcal{B}_i)$, <u>for some</u> $f \in C^\infty(R^i,R)$.

(*) The preparation of this paper was supported by the National Research Council under Grant A8731.

The proof follows directly from proposition 4.2 and lemma 4.3 in §4. In this section it is also shown that the operators $\mathcal{B}_k$, $1 \leq k \leq i$ of theorem 1.1 may be polynomial operators with polynomial coefficients if $M = R^m$ $N = R^n$ and G acts linearly, Theorem 4.4. Both proofs depend on recent results of G.W. Schwarz on invariant functions, see [9]. The main result in §3 is Theorem 3.1 stating that the invariant functions on the jet bundle $J^k(M,N)$ are completely determined by those on the isotropy subgroup of some $x_o \in M$. This extends the applicability of the above theorems to the case where the stabilizer of some $x_o \in M$ is compact. The various actions of G are treated in §2.

I would like to thank Ivan Kupka for some very helpful discussions.

2. THE ACTIONS OF G

Let M and N be smooth manifolds, dim M = m, dim N = n and G a Lie group acting smoothly on M. It is always assumed that G acts from the left. The diffeomorphism of M corresponding to $g \in G$ is denoted by ϕ_g, e is the neutral element of G. The action of G on M induces an action on $C^\infty(M,N)$, by

$$g.f \equiv \phi_g^{-1*} f , \tag{1}$$

where $g \in G$, $f \in C^\infty(M,N)$ and $\phi_g^{-1*} f = f \circ \phi_g^{-1}$.

Let X,Y and Z be smooth manifolds. $J^k(X,Y)$ is the jet bundle from X to Y. If $\sigma \in J^k(X,Y)$ is represented by f and $\tau \in J^k(Y,Z)$ is represented by h while $\beta(\sigma) = \alpha(\tau)$, where α is the source map and β is the target map, then the composition of τ and σ is defined by

(2) $$\tau \circ \sigma = j^k_{\alpha(\sigma)}(h \circ f) .$$

Let $\dim X = \dim Z$ and $\psi: Z \to X$ a diffeomorphism then $\psi^*: J^k(X,Y) \to J^k(Z,Y)$ is defined by

(3) $$\psi^*\sigma = j^k_{\psi^{-1}(\alpha(\sigma))} f \circ \psi .$$

The action of G on M lifts to a smooth action on $J^k(M,N)$ by

(4) $$g.\sigma = \phi_g^{-1*}\sigma .$$

If $\sigma \in J^k_{x,y}(M,N)$ is represented by f then ϕ_g^{-1*} is the map: $\sigma \to$ the equivalence class of $f \circ \phi_g^{-1}$ in $J^k_{\phi_g(x),y}(M,N)$. It is easily shown that (4) indeed defines an action and that the map: $(g,\sigma) \to g.\sigma$ from $G \times J^k(M,N)$ onto $J^k(M,N)$ is smooth. Putting

(5) $$\xi_g = j^k_{\phi_g(x)}(\phi_g^{-1}) ,$$

we find from (2), (3) and (4) that

(6) $$g.\sigma = \sigma \circ \xi_g .$$

The action of G commutes with j^k in the sense that for $f \in C^\infty(M,N)$

(7) $$g.j^k_x f = j^k_{\phi_g(x)} g.f .$$

Let $P^k_{m,n}$ denote the vector space of polynomial mappings from R^m into R^n of degree less than or equal to k. If f is a smooth mapping from R^m into R^n then $t^k_x f$ denotes the Taylor series of f at x. Let M and N be open sets in R^m and R^n, respectively.

The canonical diffeomorphism of $J^k(M,N)$ onto $M \times P^k_{m,n}$ is denoted by γ. If σ is represented by f then $\gamma(\sigma) = (\alpha(\sigma), t^k_{\alpha(\sigma)}f)$. As $P^k_{m,n}$ is isomorphic to a Euclidean space R^s, there is another canonical diffeomorphism δ of $J^k(M,N)$ onto $M \times R^s$ e.g. if $\sigma \in J^2(R^2,R)$ is represented by f and $\alpha(\sigma) = x$ then δ and $e^2_x f \in R^6$ are defined by $\delta(\sigma) = \delta(j^2_x f) = (x, e^2_x f)$, where $e^2_x f = (f(x), \partial f(x)/\partial x_1, \partial f(x)/\partial x_2, \partial^2 f(x)/\partial x_1^2, \partial^2 f(x)/\partial x_1 \partial x_2, \partial^2 f(x)/\partial x_2^2)$.

The action of G on $J^k(M,N)$, M open in R^m, N open in R^n, induces actions on $M \times P^k_{m,n}$ and on $M \times R^s$ by the commutative diagram:

$$\begin{array}{ccccc}
M \times P^k_{m,n} & \xleftarrow{\ \gamma\ } & J^k(M,N) & \xrightarrow{\ \delta\ } & M \times R^s \\
\uparrow & & \uparrow & & \uparrow \\
G \times (M \times P^k_{m,n}) & \xleftarrow{\ id \times \gamma\ } & G \times J^k(M,N) & \xrightarrow{\ id \times \delta\ } & G \times (M \times R^s)
\end{array}$$

The vertical arrows denote the actions of G.

If $\sigma \in J^k(M,N)$ is represented by f and $\alpha(\sigma) = x$ then the relation between σ and $t^k_x f$ or $(x, f(x), df(x), \ldots, d^k f(x))$ as well as that between σ and $e^k_x f$ is denoted by $\sim$.

2.1 LEMMA. Let U, V and W be open sets in R^ℓ, R^m and R^n, respectively. $\eta \in J^k(U,V)$, $\nu \in J^k(V,W)$, with $\eta = j^k_{x_0} f$, $\nu = j^k_{f(x_0)} h$. Then

$$\nu \circ \eta \sim (t^k_{f(x_0)} h \circ t^k_{x_0} f)_k \, , \tag{8}$$

where $_k$ denotes reduction of the composed polynomials to order k.

PROOF. $f = t^k_{x_0} f + O \| x - x_0 \|^{k+1}$, $h = t^k_{f(x_0)} h + O \| x - f(x_0) \|^{k+1}$. A straightforward computation shows that $h \circ f = (t^k_{f(x_0)}(h) o t^k_x f) +$

$O \| x-x_o \|^k$ so that $j^k_{x_o}(h \circ f) = (t^k_{f(x_o)} h \circ t^k_{x_o} f)_k$.

2.2 COROLLARY. <u>If</u> M <u>is an open set of</u> R^m <u>and</u> N <u>one of</u> R^n <u>then for</u> $g \cdot \sigma$, <u>as in</u> (4) <u>we have</u>

$$(9) \qquad g \cdot \sigma \sim (t^k_x f \circ t^k_{\phi_g(x)} \phi_g^{-1})_k.$$

2.3 LEMMA. <u>Let</u> M <u>be open in</u> R^m <u>and</u> G <u>a subgroup of</u> GL(m), <u>then for</u> $A \in G$

$$(10) \qquad A \cdot \sigma \sim (Aa, f(a), A^{-1*} df(a), \ldots, A^{-1*} d^k f(a)),$$

<u>where</u> $f \in C^\infty(M, R^n)$ <u>represents</u> σ <u>and</u> $\alpha(\sigma) = a$.

PROOF. By corollary 2.2, $A \cdot \sigma \sim \sum_{i=0}^{k} d^i f(a) (A^{-1}x - a)^i$ which equals $\sum_{i=0}^{k} A^{-1*} d^i f(a) (x - Aa)^i$.

3. INVARIANT FUNCTIONS FOR TRANSITIVE ACTION

In this section it is assumed that the action of G on M is transitive. M,N and L are smooth manifolds. For $x \in M$, S_x denotes the stabilizer of x in G and r_x the orbital map: $G \to M$, $r_x(g) = g \cdot x$. $C^\infty(M,N)^G = \{f \in C^\infty(M,N) : g \cdot f = f, \forall g \in G\}$.

For some $x_o \in M$, let

$$(1) \qquad u \in C^\infty(J^k_{x_o}(M,N), L)^{S_{x_o}}$$

be given, then $\tilde{u} : J^k(M,N) \to L$ is defined by

$$(2) \qquad \tilde{u}(\sigma) = u(g^{-1} \cdot \sigma), \text{ with } g \in G \text{ satisfying}$$

$$(3) \qquad g^{-1} \cdot \alpha(\sigma) = x_o.$$

If $h \in G$ also satisfies (3) then $h^{-1}g \in S_{x_o}$ so $u(h^{-1}.\sigma) = (h^{-1}g).u(h^{-1}.\sigma) = u(g^{-1}.\sigma)$ hence $\widetilde{u}$ is well defined.

The function of $\widetilde{u}$ is invariant under the action of G, for let $h \in G$, $\sigma \in J^k(M,N)$, $g \in G$ satisfying (3), then $h.\widetilde{u}(\sigma)=\widetilde{u}(h^{-1}.\sigma)= u(g^{-1}hh^{-1}).\sigma) = \widetilde{u}(\sigma)$.

3.1 THEOREM. Let the topology of G admit a countable base, $x_o \in M$, then

(4) $$\Phi \in C^\infty(J^k(M,N),L)^G$$

if and only if there exists a

(5) $$v \in C^\infty(J^k_{x_o}(M,N),L)^{S_{x_o}},$$

such that $\Phi=\widetilde{v}$.

PROOF. Let v satisfy (5). By lemma 3.2 below, r_{x_o} is a submersion. Hence there exists a smooth section s_{x_o} of the orbital map r_{x_o}. As $s_{x_o}(\alpha(\sigma)).x_o = r_{x_o}(s_{x_o}(\alpha(\sigma))) = \alpha(\sigma)$, $g=s_{x_o}(\alpha(\sigma))$ satisfies (3). Therefore $\widetilde{u}$ may be defined by $\widetilde{u}(\sigma)=u(s_{x_o}(\alpha(\sigma))^{-1}.\sigma)$. Consequently, $\widetilde{u}$ is smooth. As was shown above, $\widetilde{u}$ is invariant under G. Putting $\Phi=\widetilde{u}$ we obtain (4).

Conversely, let Φ satisfy (4). Then let v be the restriction of Φ to $J^k_x(M,N)$. Clearly v satisfies (5). For $\sigma \in J^k(M,N)$, $g \in G$ satisfying (3), we have $\widetilde{v}(\sigma) = v(g^{-1}.\sigma)= \Phi(g^{-1}.\sigma) = \Phi(\sigma)$ as Φ is invariant.

3.2 LEMMA. If the topology of G is second countable then, for each x in M, the orbital map r_x is a submersion.

PROOF. Put dim $G=\ell$, dim $M=m$. The mapping r_x is a subimmersion of constant rank k, ([5], XVI,10). Assume that $k<m$.

At any point of G, r_x is locally equivalent to the map: $(x_1,\ldots x_\ell) \to (x_1,\ldots,x_k,0,\ldots 0)$ of R^ℓ into R^m, ([5], XVI,7). Hence, for all $g \in G$ we may choose an open neighbourhood U_g of g and a chart (V_g,ψ) of $r_x(g)$ such that ψ maps $r_x(U_g)$ into a k-dimensional subspace of R^m. Consequently, the interior of $\overline{r_x(U_g)}$ is empty.

Let $\{O_\nu\}$ be a countable base for the topology of G. Put $\{O_\nu\}' = \{O_\nu : \exists\, U_g$ such that $O_\nu \subset U_g\}$ then, for all O_ν in $\{O_\nu\}'$, the interior of $\overline{r_x(O_\nu)}$ is empty. As $\{O_\nu\}'$ still covers G, $r_x(G)$ is the countable union of nowhere dense sets. From Baire's theorem it follows that $r_x(G) \neq M$, which contradicts the transitivity. Consequently k=m.

4. DIFFERENTIAL INVARIANTS

Let F be a given smooth function on $J^k(M,N)$. The corresponding differential operator $\mathcal{F} : C^\infty(M,N) \to C^\infty(M,R)$ is defined by

$$\mathcal{F}\, f(x) = F(j^k_x f). \tag{1}$$

4.1 LEMMA. <u>$\mathcal{F}$ is</u> G-<u>equivariant if and only if</u> F <u>is</u> G-<u>invariant</u>.

PROOF. If F is G-invariant then $\mathcal{F}(g\cdot f)(x) = F(j^k_x g\cdot f) = F(g\cdot j^k_{\phi^{-1}(x)} f)$, by (2.7), so $\mathcal{F}(g.f)(x) = g^{-1}\cdot F(j^k_{\phi_g^{-1}(x)} f) = F(j^k_{\phi_g^{-1}(x)} f) = \mathcal{F} f\,(\phi_g^{-1}(x)) = (g\cdot\mathcal{F} f\,)\,(x)$. Conversely, if $\mathcal{F}$ is G-equivariant and $\sigma \in J^k(M,N)$ is represented by f, $\alpha(\sigma)=x$, then $g\cdot F(\sigma) = g\cdot F(j^k_x f) = F(j^k_{\phi_g^{-1}(x)} g^{-1}\cdot f) = \mathcal{F}(g^{-1}\cdot f)\;(\phi_g^{-1}(x)) =$

$(g^{-1}\cdot\mathcal{F} f)(\phi_g^{-1}(x)) = \mathcal{F} f(x) = F(\sigma).$

As in §1, $\mathcal{D}_k^G$ denotes the set of smooth G-equivariant differential operators of $C^\infty(M,N)$ into $C^\infty(M,R)$. The next proposition follows directly from theorem 2 in [9] by Schwarz and the above lemma.

4.2 PROPOSITION. _If_ G _is a compact Lie group acting smoothly on a_ C^∞_-manifold_ M _such that the orbit structure of the induced action on_ $J^k(M,N)$ _is of finite type then, for all_ $k \geq 1$, _there exist_ $\mathcal{B}_1,\ldots,\mathcal{B}_i \in \mathcal{D}_k^G$ _such that_ $\mathcal{F} \in \mathcal{D}_k^G$ _if and only if_ $\mathcal{F}=f(\mathcal{B}_1,\ldots,\mathcal{B}_i)$, _for some_ $f \in C^\infty(R^i,R)$.

Regarding the orbit structure on the jet bundle the proposition is amplified by lemma 4.4.

In the proof of the next lemma some simple properties of the orbit type are needed. Let G be a smooth Lie group acting on smooth manifolds X_1 and X_2.

(1) If $X_1 \subset X_2$ and the action on X_2 is of finite type then so is that on X_1, if X_1 is invariant.

(2) If the action on X_1 is of finite type while that on X_2 is trivial then the action on $X_1 \times X_2$ is of finite type.

(3) If the actions on X_1 and X_2 are of finite type then so is that on $X_1 \cup X_2$.

(4) If $\psi: X_1 \to X_2$ is a diffeomorphism commuting with the action and the orbit structure on X_1 is of finite type then so is that on X_2.

4.3 LEMMA. _If_ G _is a compact Lie group acting on a smooth compact manifold_ M _and_ N _is a smooth manifold, countable at_ _infinity, then the induced action on_ $J^k(M,N)$ _has only a finite number of orbit types._

PROOF. First consider the case $N=R^n$ then $J^k(M,R^n) \xrightarrow{\alpha} M$ is a vector bundle. Put $\alpha^{-1}(x) = J_x^k(M,R^n)$. For each $g \in G$, the restriction to $J_x^k(M,R^n)$ of the mapping $\sigma \to g\cdot\sigma$ on $J^k(M,R^n)$ is a vector space isomorphism onto $J^k_{\phi_g(x)}(M,R^n)$, see (2.7), so G acts by automorphism on $J^k(M,R^n) \xrightarrow{\alpha} M$. Using a Haar measure one easily constructs a G-invariant scalar product (,) with corresponding norm $\| \ \|$ on $J^k(M,R^n) \xrightarrow{\alpha} M$.

The sphere bundle $S(J^k(M,R^n)) = \{\sigma \in J^k(M,R^n) : \|\sigma\| = 1\}$ is an invariant submanifold of $J^k(M,R^n)$. As $S(J^k(M,R^n))$ is compact its orbit structure is of finite type.

Let $\overset{\circ}{J}{}^k(M,R^n)$ denote the zero-section of $J^k(M,R^n)$ and $\tilde{J}^k(M,R^n)$ the complement of $\overset{\circ}{J}{}^k(M,R)$ in $J^k(M,R^n)$. $\tilde{J}^k(M,R^n) = S(J^k(M,R^n)) \times R_+^*$, ($R_+^*$ is the multiplicative group of positive reals). The action on R_+^* is trivial so the action on $\tilde{J}^k(M,R^n)$ is of finite orbit type by property (2). The diffeomorphism $\alpha: \overset{\circ}{J}{}^k(M,R^n) \to M$ commutes with the action of G. As M is compact its orbit type is finite and by property (4) the same is true for $\overset{\circ}{J}{}^k(M,R^n)$. Finally, it follows from property (3) that the orbit structure of $J^k(M,R^n)$ is of finite type.

N, being countable at infinity, may be embedded in some R^ℓ which gives an equivariant embedding of $J^k(M,N)$ in $J^k(M,R^\ell)$. As the action of G on $J^k(M,R^\ell)$ has an orbit structure of finite type, the same is true for that on $J^k(M,N)$. This completes the proof.

The action of G on elements $(x, e_x^k f) \in R^m \times R^s$ canonically induced by that on R^m was determined in §2. If G is compact and acts orthogonally on $R^m \times R^s$ then the R-algebra of G-invariant polynomial functions from $R^m \times R^s$ into R is finitely generated. Let $P_1, \ldots, P_i$ be generators. The corresponding G-equivariant operators are denoted by $\mathcal{P}_1, \ldots, \mathcal{P}_i$. For $f \in C^\infty(R^m, R^n)$,

$$\mathcal{P}_\ell f(x) \equiv P_\ell(x, e_c^k f), \quad 1 \le \ell \le i. \tag{2}$$

Let $\mathcal{D}_k^G$ be defined as above with $M=R^m$, $N=R^n$. By Weyl's unitary trick G leaves some scalar product invariant so Schwarz' theorem 1, [9] can be applied to obtain

4.4 THEOREM. *If* G *is a compact Lie group acting linearly on* R^m *then for all* $k \geq 1$, *there exist polynomial operators with polynomial coefficients*, $\mathcal{P}_1, \ldots, \mathcal{P}_i \in \mathcal{D}_k^G$, *which may be constructed as in* (2), *such that* $\mathcal{F} \in \mathcal{D}_k^G$ *if and only if* $F = f(\mathcal{P}_1, \ldots, \mathcal{P}_i)$, *for some* $f \in C^\infty(R^i, R)$.

BIBLIOGRAPHY

1. A. Borel, Seminar on Transformation Groups, *Annals of Mathematical Studies*, 46, Princeton University Press, Princeton, New Jersey, 1960.

2. N. Bourbaki, *Lie groups and Lie algebras*, part I, Addison-Wesley, 1975.

3. C. Buttin and P. Molino, Théorème general d'equivalence pour les pseudogroupes de Lie plats transitifs, *J. Differential Geometry* 9 (1974) 347-354.

4. J. Diedonne, Treatise on analyse, vol. II, Academic Press 1976.

5. J. Diedonne, Elements d'analyse, vol. III, Gauthier-Villars, 1974.

6. A. Kumpera, Invariants differentiels d'un pseudogroupe de Lie. I, II. J. Differential Geometry, 10 (1975) 289-345, 347-416.

7. S. Lie, Allgemeine Untersuchungen über Differentialgleichungen die eine kontinuierliche, endliche Gruppe gestatten, Math. Ann. 25 (1885) 77-151; Gesam. Abh. Bd. VI, 139-223.

8. L.V. Ovsjannikov, On differential invariants of a local Lie group of transformations, Soviet Math. Dokl. 16 (1975) 772- 776.

9. G.W. Schwarz, Smooth functions invariant under the action of a compact Lie group, Topology 14 (1975) 63-68.

10. A. Tresse, Sur les invariants differentiels des groupes continus de transformations, Thèse, Paris, 1893; Acta Math. 18 (1894) 1-88.

ADDENDUM. In theorem 1.1, the requirement that M be compact may be replaced by the weaker condition that the action of G on M is of finite orbit type. The proof was presented at the conference and will be published elsewhere.

UNIVERSITY OF TORONTO,
TORONTO,
ONTARIO,
M5S 1A1.
CANADA.

EQUIVARIANT REGULAR NEIGHBOURHOODS

ALLAN L. EDMONDS

1. INTRODUCTION

In this paper L. Siebenmann's theory of open regular neighbourhoods [7,8,9] is generalized to the equivariant case. As suggested in [7; §4] and [8; §6] the basic definitions and axiomatic properties of such neighbourhoods set out in §2 of this paper follow exactly the outline of the non-eqivariant case. The main non-trivial result here about equivariant regular neighbourhoods per se is an existence theorem based upon the fundamental existence theorem of [9]. The idea is to give conditions which guarantee that a subspace of a G-space admits equivariant regular neighbourhoods if it admits ordinary regular neighbourhoods. See Theorem 3.4 for a precise statement.

The nicest immediate applications of the general theory are to semifree finite group actions with isolated fixed points on manifolds. In this case the existence theorem referred to above shows that (under suitable dimension restrictions) each fixed point is contained in arbitrarily small invariant open disk neighbourhoods.

In §§4 and 5 further applications are made. These could have been handled directly by ad hoc arguments in each case. But it is enlightening and no more difficult in the end to develop the unifying general theory of equivariant regular neighbourhoods first.

In §4 it is shown (with some dimension restrictions) that the space of all actions of a finite group on a compact manifold (with the compact-open topology) is locally contractible at each semifree action with finite fixed point set. This eliminates some extraneous assumptions from a similar result in [2; (2.8)].

In §5 the equivariant homeomorphism classes of semifree actions of a finite group G on the (n+1)-sphere S^{n+1} with fixed point set the standard S^0 are classified (when $n \geq 5$) in terms of an exact sequence involving the set of equivariant homeomorphism classes of free G actions on S^n, the Whitehead group Wh(G), and the reduced projective class group $\widetilde{K}_0(ZG)$. This uses the results of §3 together with the topological version of the results of Siebenmann's thesis [5]. See Theorem 5.1 for the precise statement of the result. Further examples are constructed to show that a fixed point of such an action on S^{n+1} need not be contained in an invariant closed disk, a situation which contrasts sharply with the smooth or piecewise linear cases.

2. SOME DEFINITIONS

Recall Siebenmann's definition of (open, isotopy) regular neighbourhoods. Let X be a closed subspace of a Hausdorff space Y. Let $V \subset U \subset Y$; one says that V is <u>compressible toward X within U</u>, abbreviated 'V↘X in U', if for every neighbourhood W of X there is an isotopy $h_t: Y \to Y$, $0 \leq t \leq 1$, such that $h_0 = 1_Y$ (the identity of Y) $h_1(V) \subset W$, and for all t h_t fixes Y - U and a neighbourhood of X (independent of t). To say that h_t is an <u>isotopy</u> means that the induced map $Y \times I \to Y \times I$ is a homeomorphism.

A neighbourhood E of X in Y is said to be regular if $E = \cup E_n$, where $E_0 \subset E_1 \subset E_2 \subset \ldots$ is a sequence of neighbourhoods of X such that for each $n \geq 0$ $E_n \searrow X$ in E_{n+1}. If X admits regular neighbourhoods it admits arbitrarily small ones. The subspace X admits regular neighbourhoods if and only if a compression axiom Comp(Y,X) ('For any neighbourhood U of X in Y there is a neighbourhood V of X in U such that $V \searrow X$ in U') holds. If E and E′ are regular neighbourhoods of X in Y then there is a slide $g_t : E \to Y$, $0 \leq t \leq 1$, fixing a neighbourhood of X (independent of t) such that g_0 is the inclusion and $g_1(E) = E'$. To say that g_t is a slide means that the associated map $E \times I \to Y \times I$ is an open embedding.

Now let G be a compact Lie group, Y be a G-space, and X be a closed invariant subspace of Y. Let $V \subset U \subset Y$ be invariant subspaces; one says that V is G-compressible toward X within U, abbreviated '$V \overset{G}{\searrow} X$ in U', if for any neighbourhood W of X in Y there is an equivariant isotopy $h_t : Y \to Y$, $0 \leq t \leq 1$, such that $h_0 = 1_Y$, $h_1(V) \subset W$, and for all t h_t fixes $Y - U$ and a neighbourhood of X (independent of t).

An invariant neighbourhood E of X in Y is said to be a G-regular neighbourhood if $E = \cup E_n$, where $E_0 \subset E_1 \subset E_2 \subset \ldots$ is a sequence of invariant neighbourhoods of X in Y such that $E_n \overset{G}{\searrow} X$ in E_{n+1} for all $n \geq 0$.

By essentially the same arguments as in the non-equivalent case the following properties hold: if X admits G-regular neighbourhoods, it admits arbitrarily small ones. The subspace X admits

G-regular neighbourhoods in Y if and only if the G-compression axiom G-Comp(Y,X) ('For any neighbourhood U of X in Y there is an invariant neighbourhood V of X in U such that $V \searrow^{G} X$ in U') holds. If E and E′ are G-regular neighbourhoods of X in Y there is a G-slide $h_t : E \to Y$, $0 \leq t \leq 1$, fixing a neighbourhood of X such that h_0 is the inclusion, and $h_1(E) = E'$.

Now let Σ be any set. A Σ-space is a space Y together with a function $Y \to \Sigma$. In application here Σ will be a set of orbit types for a compact Lie G and Σ-spaces will arise naturally as orbit spaces for G-actions. A map $Y_1 \to Y_2$ of Σ-spaces is a Σ-map if the diagram

$$\begin{array}{ccc} Y_1 & \longrightarrow & Y_2 \\ & \searrow \quad \swarrow & \\ & \Sigma & \end{array}$$

commutes. A subspace of a Σ-space is naturally a Σ-space such that the inclusion is a Σ-map.

If Y is a Σ-space and X is a closed subspace we obtain obvious notions of Σ-compressibility '$V \searrow^{\Sigma} X$ in U' and of Σ-regular neighbourhoods. Also Σ-regular neighbourhoods exist if and only if a suitable axiom Σ-Comp(Y,X) holds and Σ-regular neighbourhoods are unique via Σ-slides fixed near X.

3. AN EXISTENCE THEOREM FOR G-REGULAR NEIGHBOURHOODS

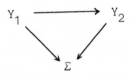

Let G be a compact Lie group and let Σ be the set of equivariant homeomorphism classes of G-orbits G/H, where H varies over the closed subgroups of G. Let Y be a metric G-space and $X \subset Y$ be a closed invariant subspace. For any $Z \subset Y$ let Z^* be the image of Z in the orbit space $Y/G = Y^*$. Now Y^* is a Σ-space, where $Y^* \to \Sigma$ simply assigns to an orbit its equivariant homeomorphism type.

THEOREM 3.1. *The subspace X admits G-regular neighbourhoods in Y if and only if X^* admits Σ-regular neighbourhoods in Y^*.*

PROOF. If $E = \cup E_n$ is a G-regular neighbourhood of X in Y then $E^* = \cup E_n^*$ is clearly a Σ-regular neighbourhood of X^* in Y^*, since equivariant compressions in Y induce Σ-compressions in Y^*.

Conversely, suppose $E^* = \cup E_n^*$ is a Σ-regular neighbourhood of X^* in Y^*. Let $E_n = \pi^{-1} E_n^*$ (where $\pi: Y \to Y^*$ is the orbit map) and $E = \pi^{-1} E^* = \cup E_n$. To show that E is a G-regular neighbourhood of X in Y it suffices to show that $E_n \searrow^G X$ in E_{n+1}, given that $E_n^* \searrow^\Sigma X^*$ in E_{n+1}^*. Let W be any invariant neighbourhood of X in Y and let $h_t^*: Y^* \to Y^*$, $0 \leq t \leq 1$, be a Σ-isotopy such that $h_0^* = 1_{Y^*}$, $h_1^*(E_n^*) \subset W^*$, and $h_t^* = 1_{Y^*}$ on $Y^* - E_{n+1}^*$ and on a small neighbourhood V^* of X^*. By (3.2) below (a relativised version of the Covering Homotopy Theorem of R. Palais) h_t^* can be lifted to a G-isotopy $h_t: Y \to Y$ such that $h_0 = 1_Y$, $h_1(E_n) \subset W$, and $h_t = 1$ on $Y - E_{n+1}$ and a neighbourhood of X contained in $\pi^{-1}(V^*)$.

For any G-space Z let π_Z denote the orbit map $Z \to Z^*$.

THEOREM 3.2 (Relative Covering Homotopy Theorem). *Let Y and Z be metric G-spaces with $A \subset Z$ a closed invariant subspace and let $h_t^*: Z^* \to Y^*$ be a Σ-homotopy. Suppose there is a G-map $f_0: Z \to Y$ such that $\pi_Y f_0 = h_0^* \pi_Z$ and suppose there is an invariant neighbourhood U of A in Z and a G-homotopy $f_t: U \to Y$ of $f_0|U$ such that $\pi_Y f_t = (h_t|U)\pi_U$. Then there is a G-homotopy $h_t: X \to Y$ such that $h_0 = f_0$, $h_t = f_t$ on a neighbourhood of A within U, and $\pi_Y h_t = h_t^* \pi_Z$.*

PROOF SKETCH. Write $Z = U \cup V$ where V is an open invariant set such that the closure $\overline{V}$ misses A. The non-relative version of the Covering Homotopy Theorem [1; II.7.3], for example, provides a G-homotopy $h'_t: V \to Y$ such that $\pi_Y h'_t = (h^*_t | V^*)\pi_V$ and $h'_0 = f_0 | V$. The argument in part (C) in the proof of [1; II.7.1] shows that f_t and h'_t can be spliced together to get the desired G-homotopy which will equal f_t on $U - V$.

COROLLARY 3.3. *Let Y be a metric G-space and X be a closed invariant subspace such that the action on Y - X has just one orbit type. Then X admits G-regular neighbourhoods in Y if and only if X* admits (ordinary) regular neighbourhoods in Y*.*

PROOF. In this situation regular neighbourhoods and Σ-regula[r] neighbourhoods of X* in Y* are the same. Apply (3.1).

The following is the main result of this section.

THEOREM 3.4. *Let G be a compact Lie group, Y be a locally compact metric G-space, and X be a compact invariant subspace of Y such that on Y - X the action has just one orbit type. Assume Y* - X* is a manifold of dimension at least six. Then X admits G-regular neighbourhoods in Y if and only if X admits ordinary regul[ar] neighbourhoods in Y.*

PROOF. Clearly the existence of G-regular neighbourhoods implies the existence of ordinary regular neighbourhoods. For the converse we shall apply the basic existence theorem of [9].

According to (3.3) it suffices to show that X* admits regula[r] neighbourhoods in Y*. Let $W = Y - X$. Then X determines a number of *ends* of W and X* likewise determines a number of ends of W*. According to [9;2.1] it suffices to show that the ends of W* at X^* are finite in number and <u>tame</u> in order to complete the proof of (3.4).

Recall that an isolated end ϵ of a manifold M is tame if there is a decreasing sequence $\{U_i\}$ of connected neighbourhoods of ϵ in M such that (i) $\operatorname{int} U_i \supset \overline{U}_{i+1}$, $\cap U_i = \emptyset$, and the system

$\pi_1 U_1 \xleftarrow{f_1} \pi_1 U_2 \xleftarrow{f_2} \pi_1 U_3 \longleftarrow \cdots$ (where f_i is induced by inclusion) is <u>essentially constant</u> - i.e. at least upon passage to a subsequence there are induced isomorphisms $\operatorname{Im} f_1 \xleftarrow{\approx} \operatorname{Im} f_2 \xleftarrow{\approx} \operatorname{Im} f_3 \xleftarrow{\approx} \cdots$ (this is independent of choices of base points); and (ii) for each i there is a finite CW complex L_i and a homotopy commutative diagram

$$\begin{array}{ccc} U_i & \hookleftarrow & U_{i+1} \\ & \nwarrow \qquad \swarrow & \\ & L_i & \end{array} .$$

See [9; 1.3].

As in [9] the actual nature of X or X^* is not important: we may assume that Y is compact and that Y is in fact the end point compactification of W and that Y^* is the end point compactification of W^*. Notice that the action of G on W extends uniquely over the end point compactification with orbit space the end point compactification of W^*.

Since X admits regular neighbourhoods under the above assumption, X consists of finitely many isolated points. Thus X^* also consists of finitely many points. By considering X^* one point at a time we may assume X^* is in fact just one point. Moreover we may further assume that X is also just one point. For let $x \in X$;

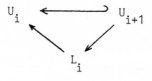

choose a compact G_x-invariant neighbourhood C of x such that $C \cap X = \{x\}$ and $C - \{x\}$ is a manifold (using handlebody theory or topological transversality [4; III.1] applied to an appropriate proper map $Y^* - X^* \to R$). Then a G_x-regular neighbourhood of $\{x\}$ in C determines a Σ-regular neighbourhood of $\{x^*\}$ in Y^*, which determines a G-regular neighbourhood of the orbit $G(x)$ in Y.

Now let $V_1^* \supset V_2^* \supset V_3^* \supset \ldots$ be a basis of compact connected neighbourhoods of X^* in Y^* such that each $\mathrm{int}\, V_n \supset \overline{V_{n+1}}$ and each $V_n^* - X^*$ is a connected manifold with compact boundary (again using topological transversality or handlebody theory [4] since $\dim Y^* - X^* \geq 6$). Let $f_n^* : \pi_1(V_{n+1}^* - X^*) \to \pi_1(V_n^* - X^*)$ be the inclusion-induced homomorphism (for some choice of base points). Let V_n be the pre-image of V_n^* in Y. Then each V_n is compact and $V_n - X$ is a connected manifold with compact boundary. Since π_1 is essentially constant at X, it follows (by passage to a subsequence if necessary) that

$$\pi_1(V_1-X) \xleftarrow{f_1} \pi_1(V_2-X) \xleftarrow{f_2} \pi_1(V_3-X) \longleftarrow \cdots$$

induces isomorphisms

$$\mathrm{Im}\, f_1 \xleftarrow{\approx} \mathrm{Im}\, f_2 \xleftarrow{\approx} \mathrm{Im}\, f_3 \xleftarrow{\approx} \cdots \quad .$$

See [5; pp.11-14].

We assert that this implies that

$$\mathrm{Im}\, f_1^* \xleftarrow{\approx} \mathrm{Im}\, f_2^* \xleftarrow{\approx} \mathrm{Im}\, f_3^* \xleftarrow{\approx} \cdots$$

is also a sequence of isomorphisms. Because $W \to W^*$ is a locally trivial fibre bundle with fibre G/H for some subgroup H [1; II.5.4], we obtain a big commutative diagram with exact columns

$$\begin{array}{ccccccc}
\pi_1 G/H & = & \pi_1 G/H & = & \pi_1 G/H & = & \cdots \\
\downarrow & & \downarrow & & \downarrow & & \\
\pi_1(V_1-X) & \longleftarrow & \pi_1(V_2-X) & \longleftarrow & \pi_1(V_3-X) & \longleftarrow & \cdots \\
\downarrow & & \downarrow & & \downarrow & & \\
\pi_1(V_1^*-X^*) & \longleftarrow & \pi_1(V_2^*-X^*) & \longleftarrow & \pi_1(V_3^*-X^*) & \longleftarrow & \cdots \\
\downarrow & & \downarrow & & \downarrow & & \\
\pi_0 G/H & = & \pi_0 G/H & = & \pi_0 G/H & = & \cdots \\
\downarrow & & \downarrow & & \downarrow & & \\
0 & & 0 & & 0 & &
\end{array}$$

This induces a commutative diagram with exact columns

$$\begin{array}{ccccccc}
\pi_1 G/H & = & \pi_1 G/H & = & \pi_1 G/H & = & \cdots \\
\downarrow & & \downarrow & & \downarrow & & \\
\mathrm{Im}\ f_1 & \overset{\approx}{\longleftarrow} & \mathrm{Im}\ f_2 & \overset{\approx}{\longleftarrow} & \mathrm{Im}\ f_3 & \overset{\approx}{\longleftarrow} & \cdots \\
\downarrow & & \downarrow & & \downarrow & & \\
\mathrm{Im}\ f_1^* & \longleftarrow & \mathrm{Im}\ f_2^* & \longleftarrow & \mathrm{Im}\ f_3^* & \longleftarrow & \cdots \\
\downarrow & & \downarrow & & \downarrow & & \\
\pi_0 G/H & = & \pi_0 G/H & = & \pi_0 G/H & = & \cdots \\
\downarrow & & \downarrow & & \downarrow & & \\
0 & & 0 & & 0 & &
\end{array}$$

A diagram chase now implies the claim. (This is essentially the argument of [6; (3.2)].)

To complete the proof of (3.4) we must find a finite CW complex L_n and a homotopy commutative diagram

$$\begin{array}{ccc}
V_n^*-X^* & \longleftarrow & V_{n+1}^*-X^* \\
& \nwarrow \quad \swarrow & \\
& L_n &
\end{array}$$

In fact we obtain a finite complex L_n and maps $r_n: L_n \to V_n^* - X^*$ and $i_n: V_n^* - X^* \to L_n$ such that $r_n i_n \simeq 1_{V_n^*-X^*}$ (i.e., we show $V_n^*-X^*$ is dominated by a finite complex).

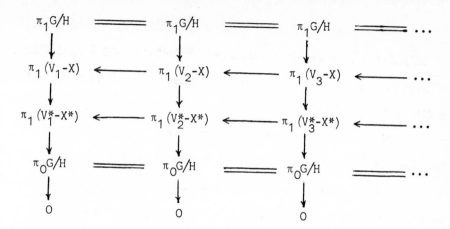

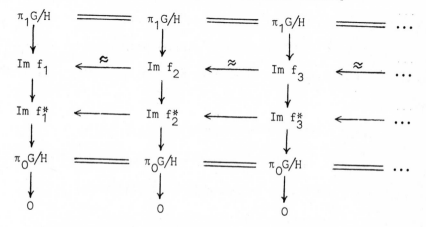

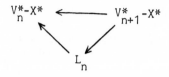

LEMMA 3.5. *Let* $F \to E \to B$ *be a fibration where* F *is a finite CW complex,* B *is a finite-dimensional* CW *complex, and* E *is a connected CW complex dominated by a finite CW complex. Then* B *is dominated by a finite CW complex.*

This will complete the proof of (3.4), for by hypothesis X has regular neighbourhoods in Y and this easily implies that V_n-X is dominated by a finite complex since we arranged that V_n-X has compact manifold frontier. Compare [9; 1.1]. Now G/H has the structure of a finite complex since it is a compact smooth manifold. Thus by (3.5) $V_n^* - X^*$ is also dominated by a finite complex.

PROOF OF (3.5). Since B is finite-dimensional it suffices to show that B satisfies the properties F_n $(n \geq 1)$ of Wall [12; §1 and Theorem F]:

F_1: $\pi_1 B$ is finitely generated.

F_2: $\pi_1 B$ is finitely-presented and for any finite 2-complex K and map $f: K \to B$ inducing an isomorphism of fundamental groups, $\pi_2(f)$ is finitely generated as a $Z[\pi_1 B]$-module.

F_n $(n \geq 3)$: F_{n-1} holds and for any finite (n-1)-complex K and (n-1)-connected map $f: K \to B$, $\pi_n(f)$ is a finitely generated $Z[\pi_1 B]$-module.

We may assume that the corresponding properties hold for the total space E.

First observe that $\pi_1 B$ is finitely presented. There is an exact sequence

$$\pi_1 F \to \pi_1 E \to \pi_1 B \to \pi_0 F \to 0 .$$

It follows easily that $\pi_1 B$ contains a finitely presented subgroup of finite index and is thus finitely presented.

Now let K be a finite (n-1)-complex, $n \geq 3$, and let $f:K \to B$ be an (n-1)-connected map. (The special case $n = 2$ is similar and is left to the reader.) Form a pullback diagram

$$\begin{array}{ccc} K' & \xrightarrow{f'} & E \\ \downarrow & & \downarrow \\ K & \xrightarrow{f} & B \end{array}$$

Then $K' \to K$ is also a fibration with fibre F. The ladder connecting the homotopy exact sequences of the two fibrations shows that $\pi_n(f') \approx \pi_n(f)$ as $Z[\pi_1 E]$-modules, where $\pi_1 E$ acts on $\pi_n(f')$ in the usual way and on $\pi_n(f)$ via the homomorphism $\pi_1 E \to \pi_1 B$.

Since E is dominated by a finite complex and K′ has the homotopy type of a finite complex [10; Proposition 1], $\pi_n(f')$ is easily seen to be finitely generated over $Z[\pi_1 E]$.

Therefore $\pi_n(f)$ is finitely generated over $Z[\pi_1 E]$. Since $\pi_1 E$ acts on $\pi_n(f)$ via $\pi_1 E \to \pi_1 B$, it follows that $\pi_n(f)$ is also finitely generated over $Z[\pi_1 B]$. Thus B satisfies F_n for all n, and being finite-dimensional is dominated by a finite complex.

REMARK. Presumably with more care the dimension restriction ≥ 6 in (3.4) can be reduced to ≥ 5. The crucial step to be modified is the application of (3.5) to the manifolds V_n-X near the end of the proof. Topological transversality for maps $Y^* - X^* \to R$ (or handlebody theory) requires dimension $\neq 5$ and so should be avoided.

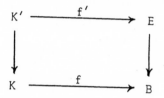

COROLLARY 3.6. *Suppose a finite group G acts semifreely on a topological m-manifold, $m \geq 6$, with only isolated fixed points. Then each fixed point is contained in arbitrarily small invariant open neighbourhoods homeomorphic to euclidean m-space.*

PROOF. The fixed point set clearly admits regular neighbourhoods. By (3.4) it admits G-regular neighbourhoods. Uniqueness of ordinary regular neighbourhoods implies the result.

In general the fixed points above are *not* contained in invariant closed disk neighbourhoods. This is the subject of §5.

4. LOCAL CONTRACTIBILITY OF SPACES OF GROUP ACTIONS

Let $\mathcal{A}(G,M)$ denote the space of all actions of a *finite* group G on a compact topological m-manifold M (endowed with the compact-open topology). We think of an action φ as an injective homomorphism $g \to \varphi^g$ of G into the homeomorphism group of M. Let $\mathcal{A}_0(G,M)$ denote the subspace of $\mathcal{A}(G,M)$ consisting of actions which are semifree and have fixed point sets which consist at most of isolated points. The following extends and clarifies Theorem 2.8 of [2].

THEOREM 4.1. *If $m \geq 6$ then $\mathcal{A}(G,M)$ is locally contractible at each φ in $\mathcal{A}_0(G,M)$. That is, there is a neighbourhood $\mathcal{U}$ of φ in $\mathcal{A}(G,M)$ such that for each $\psi \in \mathcal{U}$ there is a level-preserving G-action θ on $M \times I$ such that $\theta|M \times 0 = \varphi$, $\theta|M \times 1 = \psi$, and θ depends continuously on ψ.*

PROOF. Let $E_0 \supset E_1 \supset E_2 \supset \ldots$ be a decreasing sequence of G-regular neighbourhoods of the fixed point set F of φ guaranteed by (3.4) such that $\cap E_n = F$ and $E_n \supset \bar{E}_{n+1}$, for all $n \geq 0$.

It follows that each E_n, $n \geq 1$, G-compresses toward F into E_{n+1} within E_{n-1}. For each $n \geq 1$ choose a φ-equivariant isotopy f_t^n: $M \to M$, $(n-1)/n \leq t \leq n/(n+1)$, such that $f^n_{(n-1)/n} = 1_M$, $f^n_{n/(n+1)}(\bar{E}_n) \subset E_{n+1}$, and for all t $f_t^n = 1_M$ on $\overline{M-E_{n-1}}$ and a neighbourhood of F. Extend f_t^n for $0 \leq t \leq 1$ by setting $f_t^n = f^n_{(n-1)/n}$ if $0 \leq t \leq (n-1)/n$ and $f_t^n = f^n_{n/(n+1)}$ if $n/(n+1) \leq t \leq 1$.

Now define f_t:$M \to M$ for

$$0 \leq t < 1 \text{ by } f_t = f_t^n\, f_t^{n-1} \quad \cdots \quad f_t^1 ,$$

where $(n-1)/n \leq t \leq n/(n+1)$. This gives a well-defined isotopy since if $t = (n-1)/n$ $f_t^n = 1_M$, so that the two possible definitions of f_t^n agree.

For ψ sufficiently near φ [2; (2.3)] provides a canonical path from ψ to some ψ' such that $\psi' = \varphi$ on $M-E_1$. Therefore we may assume that $\psi = \varphi$ on $M-E_1$. Now define a path θ_t, $0 \leq t \leq 1$, in $\mathcal{A}(G,M)$ by setting $\theta_t^g = f_t \psi^g f_t^{-1}$ (if $0 \leq t < 1$) and $\theta_1 = \varphi$. One verifies that θ_t is a well-defined path as follows: let $x \in M$. If $x \in M - E_0$, $f_t^{-1}(x) = x$ and $\psi^g(x) = \varphi^g(x)$ so $\theta_t = \varphi = \psi$ there. If $x \in E_0 - F$, then there is a parameter s such that $f_t^{-1}(x) \in E_0 - E_1$ for $t \geq s$. Thus for $s \leq t < 1$,

$$\theta_t^g(x) = f_t \psi^g f_t^{-1}(x) = f_t \varphi^g f_t^{-1}(x) = \varphi^g(x) .$$

Finally if $x \in F$, then $\psi^g(x) \in E_1$; so that $f_t \psi^g(x) \to x$ as $t \to 1$ since F consists only of isolated points. Thus

$$\theta_t^g(x) = f_t \psi^g f_t^{-1}(x) = f_t \psi^g(x) \to x \qquad \text{as } t \to 1.$$

Thus θ_t determines a canonical path in $\mathcal{A}(G,M)$ depending continuously on ψ near φ, beginning at ψ and ending at φ.

REMARK. The above argument also shows that the space of free actions of G on an open manifold with finitely many ends each of which is tame is locally contractible. A modified version of the argument with open regular neighbourhoods shows that the full homeomorphism group of such a manifold is also locally contractible. The latter observation generalizes [3; Corollary 6.1] where a similar result was proved for manifolds which are interiors of compact manifolds.

5. SEMIFREE ACTIONS ON SPHERES WITH TWO FIXED POINTS

Let G be a finite group and let $\mathcal{A}_0^*(G,S^{n+1})$ denote the set of equivariant homeomorphism classes rel S^0 of semifree G actions on the standard (n+1)-sphere S^{n+1} which have fixed point set the standard S^0. Let $\mathcal{F}\mathcal{A}^*(G,S^n)$ denote the set of equivariant homeomorphism classes of free G-actions on S^n. ($\mathcal{F}\mathcal{A}^*$ is the space of free actions $\mathcal{F}\mathcal{A}$ modulo the action of the homeomorphism group.)

THEOREM 5.1. <u>If</u> $n \geq 5$ <u>there is an exact sequence of sets</u>

$$0 \to \mathcal{F}\mathcal{A}^*(G,S^n)/\mathrm{Wh}(G) \xrightarrow{S} \mathcal{A}_0^*(G,S^{n+1}) \xrightarrow{\sigma} \tilde{K}_0(ZG).$$

PROOF SKETCH AND EXPLANATION. The Whitehead group Wh(G) acts on $\mathcal{F}\mathcal{A}^*(G,S^n)$ as follows: if $\tau \in \mathrm{Wh}(G)$ and $\varphi \in \mathcal{F}\mathcal{A}(G,S^n)$ let W be the h-cobordism starting with S^n/φ and having torsion $\tau(W,S^n/\varphi) = \tau$. Let N be the other boundary component of W. The universal covering $\tilde{W}$ is an h-cobordism of S^n on which G acts by covering translations. By the h-cobordism theorem $\tilde{N} \cong S^n$. Then $\tau^*[\varphi]$ is represented by $\tilde{N}$ with its induced G-action.

The function $S:\mathcal{F}\mathcal{A}^*(G,S^n) \to \mathcal{A}_0^*(G,S^{n+1})$ is just suspension. The fact that suspension induces a well-defined injection (still called S)

$$\mathcal{F}\mathcal{A}^*(G,S^n)/Wh(G) \to \mathcal{A}_0^*(G,S^{n+1})$$

follows by applying to orbit spaces the standard fact that two manifolds have homeomorphic suspensions if and only if they are h-cobordant.

Now suppose φ is an action representing an element $[\varphi]$ in $\mathcal{A}_0^*(G,S^{n+1})$. Let $W = S^{n+1} - S^0/\varphi$. By (3.4) S^0 admits G-regular neighbourhoods in S^{n+1} and, equivalently, the ends of W are tame. Let ϵ be the end of W corresponding to the 'north pole' of S^{n+1}. Let $\sigma[\varphi]$ be Siebenmann's obstruction in the reduced projective class group $\tilde{K}_0(ZG)$ to collaring the end ϵ [5].

Clearly the ends of W are collared if φ is a suspension, so that $\sigma S = 0$. Conversely, suppose $\sigma[\varphi] = 0$. Then by [5] the end ϵ of $W = S^{n+1} - S^0/\varphi$ is collarable. In fact W splits as $N \times R$ where N is an n-manifold. To see this let $N \times [0,\infty) \to W$ be a collaring of ϵ. Let $V = W - N \times (0,\infty)$. One easily argues that the inclusion $N \times 0 \hookrightarrow V$ is a homotopy equivalence and that π_1 is essentially constant at ∞ in V. Thus the hypotheses of the open collar theorem of [6] are satisfied, so that $V = N \times (-\infty,0]$ and $W = N \times R$. Hence φ is a suspension.

It is difficult to describe precisely the image of σ in $\tilde{K}_0(ZG)$. In general σ is neither surjective nor zero. For the remainder of this section assume n is odd (since if n is even Smith theory implies that $G = Z/2$ and $\tilde{K}_0(ZZ/2) = 0$).

LEMMA 5.2. *Let* $[\varphi] \in \mathcal{A}_0^*(G, S^{n+1})$ *and* $\alpha \in \tilde{K}_0(ZG)$. *Then there is* $[\psi] \in \mathcal{A}_0^*(G, S^{n+1})$ *such that* $\sigma[\psi] = \sigma[\varphi] + \alpha + \bar{\alpha}$.

Here $\bar{\alpha}$ denotes the image of α under the standard involution of $\tilde{K}_0(ZG)$ induced by inversion in G.

PROOF SKETCH. Let P be a finitely generated projective ZG-module representing α. There is a short exact sequence

$$0 \to F_2 \to F_1 \to P \to 0$$

where F_1 and F_2 are (infinitely generated) free modules.

Let $W = S^{n+1} - S^0/\varphi$ and let V be an (n-2)-neighbourhood of the north pole end ε of W - that is, V and ∂V are connected manifolds $\pi_1(\partial V) \approx \pi_1 V \approx \pi_1 W \approx G$, $H_i(\tilde{V}, \partial\tilde{V}) = 0$ for $i \neq n-1$, and $H_{n-1}(\tilde{V}, \partial\tilde{V})$ is a finitely generated projective ZG-module representing $\sigma[\varphi]$ in $\tilde{K}_0(ZG)$.

Form the infinite, interior, locally-finite connected sum of V with copies of $S^2 \times S^{n-1}$, one for each element of a basis for the free ZG-module F_1. Let V_1 be the resulting manifold. Then

$$H_i(\tilde{V}_1, \partial\tilde{V}_1) \approx H_i(\tilde{V}, \partial\tilde{V}) \qquad \text{for } i \neq 2, n-1;$$

$$H_2(\tilde{V}_1, \partial\tilde{V}_1) \approx F_1; \text{ and}$$

$$H_{n-1}(\tilde{V}_1, \partial\tilde{V}_1) \approx H_{n-1}(\tilde{V}, \partial\tilde{V}) \oplus F_1.$$

Let V_2 be the result of killing the image of F_2 in $F_1 \approx H_2(\tilde{V}_1, \partial\tilde{V}_1)$ by locally finite surgery on a ZG-basis represented by embedded 2-spheres. (These classes are clearly spherical and represented by locally finite embedded spheres with trivial normal bundles.) Then $H_i(\tilde{V}_2, \partial\tilde{V}_2) \approx H_i(\tilde{V}, \partial\tilde{V})$ for $i \neq 2$, n-1; $H_2(\tilde{V}_2, \partial\tilde{V}_2) \approx$ and $H_{n-1}(\tilde{V}_2, \partial\tilde{V}_2) \approx H_{n-1}(\tilde{V}, \partial\tilde{V}) \oplus \bar{P}$. Compare [5; Chapter XI] and [13 Theorem 1.5].

Let ϵ_2 be the end of V_2 (corresponding to the end ϵ of W). Evidently ϵ_2 is tame and the collaring obstruction for ϵ_2 can be identified with the class of $P \oplus (H_{n-1}(\tilde{V},\partial\tilde{V}) \oplus \bar{P})$, namely $\sigma[\varphi] + \alpha + \bar{\alpha}$ (recall that n is odd).

Let U be an open regular neighbourhood of ϵ_2 in V_2. Notice that the composition of the operations $\tilde{V} => \tilde{V}_1$ and $\tilde{V}_1 => \tilde{V}_2$ does not change the (tame, collared) end of $\tilde{V}$ - namely $S^n \times R$. Thus $\tilde{U} \cong S^n \times R$. The action of G on $\tilde{U}$ by covering translations induces the desired action ψ on the end point compactification S^{n+1} of $\tilde{U}$.

COROLLARY 5.3. <u>If</u> G <u>acts freely on</u> S^n <u>then</u>

$$\operatorname{Im} \sigma \supset \{\alpha + \bar{\alpha}: \ \alpha \in \tilde{K}_0(ZG)\}$$

At least in some cases there are further restrictions on the possible values of $\sigma[\varphi]$. One easily modifies the proof of the duality theorem of Siebenmann [5; 11.1] to show that

$$\sigma(\epsilon_-) = (-1)^n \overline{\sigma(\epsilon_+)} = \overline{-\sigma(\epsilon_+)}$$

where ϵ_- and ϵ_+ denote the two ends of W ($\epsilon_+ = \epsilon$). Also, by the sum theorem for finiteness obstructions, up to sign the Wall obstruction $\sigma W = \sigma(\epsilon_-) + \sigma(\epsilon_+) = \epsilon(\sigma_+) - \overline{\sigma(\epsilon_+)}$.

Now restrict to the case that G = Z/p, a finite cyclic group of order p. In this case $\sigma W = 0$, since σW is an invariant of homotopy type and the finite lens spaces exhaust all possible homotopy types [11; 2.4]. Therefore $\sigma(\epsilon_+) = \overline{\sigma(\epsilon_+)}$. Since $\overline{\alpha + \bar{\alpha}} = \alpha + \bar{\alpha}$, these facts can be summarized in the following result.

COROLLARY 5.4. If G = Z/p then

$$\{\alpha + \bar{\alpha}: \alpha \in \tilde{K}_0(ZG)\} \subset \operatorname{Im} \sigma \subset \{\alpha \in \tilde{K}_0(ZG): \alpha = \bar{\alpha}\} .$$

To describe Im σ any more precisely is quite similar to determining which Wall obstructions can arise for Poincaré complexes [13].

REMARK. Most of the discussion in this section applies equally well to the analysis of free end-preserving actions on $M \times R$ where M is any compact manifold.

REFERENCES

1. G.E. Bredon, Introduction to Compact Transformation Groups, Academic Press, New York and London, 1972.

2. A.L. Edmonds, Local connectivity of spaces of group actions, Quart. J. Math. Oxford (2), 27 (1976), 71-84.

3. R.D. Edwards and R.C. Kirby, Deformations of spaces of embeddings, Ann. of Math. 93 (1971), 63-88.

4. R.C. Kirby and L.C. Siebenmann, Essays on topological manifolds smoothings and triangulations, to appear.

5. L.C. Siebenmann, The obstruction to finding a boundary for an open manifold of dimension greater than five, Princeton Thesis, 1965.

6. L.C. Siebenmann, On detecting open collars, Trans. Amer. Math. Soc. 142 (1969), 201-227.

7. L.C. Siebenmann, Regular open neighbourhoods, General Topology and its Applications 3 (1973), 51-61.

8. L.C. Siebenmann, L. Guillou and H. Hahl, Les voisinages ouverts réguliers, Ann. scient. Éc. Norm. Sup. 6 (1973), 253-294

9. L.C. Siebenmann, L. Guillou and H.Hahl, Les voisinages ouverts réguliers: criteres homotopique d'existence, Ann. scient. Éc. Norm. Sup. 7 (1974), 431-462.

10. J.D. Stasheff, A classification theorem for fibre spaces, Topology 2 (1963), 239-246.

11. C.B. Thomas and C.T.C. Wall, The topological spherical space form problem I, Comp. Math. 23 (1971), 101-114.

12. C.T.C. Wall, Finiteness conditions for CW-complexes, Ann. of Math. 81 (1965), 56-69.

13. C.T.C. Wall, Poincaré complexes: I, Ann. of Math. 86 (1967), 213-245.

CORNELL UNIVERSITY

ITHACA, N.Y. 14853

U.S.A.

V. GIAMBALVO

1. INTRODUCTION

Recall a Spin structure on a (compact, oriented, C^∞) n-manifold X is a principal Spin(n) bundle S over X, together with a covering map $\pi : S \to P_\tau$ where P_τ is the principal SO(n) bundle associated to the tangent bundle of X, such that the following diagram commutes.

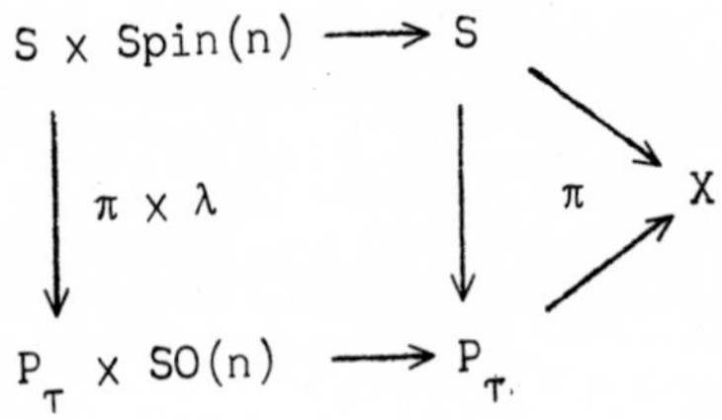

where λ is the standard double covering, and the horizontal maps are the right action on the principal bundle. A Spin structure preserving involution on (S, π, X) is a pair of involutions T: X → X and $\tilde{T}$: S → S such that $\pi\tilde{T} = dT$ and $(\tilde{T}x)a = \tilde{T}(xa)$ for any a ε Spin(n). Then the orbit space X/T has an induced Spin structure. It follows (see [4,6]) that the bordism group $\Omega_n^{Spin}(Z/2, \text{free})$ of free, Spin structure preserving involutions is isomorphic to the group $\Omega_n^{Spin}(BZ/2)$ of Spin manifolds with maps into BZ/2.

In [2] Anderson, Brown and Peterson showed that KO-Pontryagin numbers and Stiefel-Whitney numbers determine Ω_*^{Spin}.

In Section 2, KO-characteristic numbers are constructed, and it is shown that these together with Stiefel-Whitney numbers distinguish the elements of $\Omega_n^{Spin}(BZ/2)$. The main tool is the determination of $\Omega_*^{Spin}(BZ/2)$ in [3]. It follows that the only elements of order > 2 in $\tilde{\Omega}_n^{Spin}(BZ/2)$ are multiples (over Ω_*^{Spin}) of $[RP^{4n+3}, \iota]$.

2. PROOF OF MAIN THEOREM

Since the Spin structures on the stable normal bundle are equivalent to those on the tangent bundle, we will work with manifolds with a Spin structure on the normal bundle. If M^n is a Spin manifold, it is KO orientable. Let $[M^n]$ be the fundamental class in $KO_n(M)$ corresponding to the fundamental class in $KO^o(M)$ of [5] via Poincaré duality. Let π^r, r a positive integer, denote the KO-Pontryagin classes defined in [2]. For any $L = (\ell_1, \ell_2, \ldots, \ell_s)$ let $\pi^L = \pi^{\ell_1} \pi^{\ell_2} \ldots \pi^{\ell_s}$ and $\pi^o = 1$. For any element $[M,f] \in \Omega_n^{Spin}(BZ/2)$ define the KO-Pontryagin number $c(\pi^L)[M,f]$ as the image of $[M]$ under the composition

$$KO_n(M) \xrightarrow{(\nu \times f)_*} KO_n(BSpin \times BZ/2) \xrightarrow{\pi^L} KO_n(BZ/2).$$

This gives a homomorphism

$$c(\pi^L) \colon \Omega_n^{Spin}(BZ/2) \to KO_n(BZ/2).$$

Replacing KO_n by $H_n(\quad, Z/2)$ and π^L by $x \in H^*(BSpin, Z/2)$ gives the Stiefel-Whitney numbers.

THEOREM. (M_1^n, f_1) _and_ (M_2^n, f_2) _represent the same element of_ $\tilde{\Omega}_n^{Spin}(BZ/2)$ _if and only if_ (M_1^n, f_1) _and_ (M_2^n, f_2) _have the same KO-Pontryagin and Stiefel-Whitney numbers._

Since both numbers are cobordism invariants, one direction is clear. To prove the other, some information about the structure of $\Omega_*^{Spin}(BZ/2)$ is needed, as well as some facts about $KO_*(RP^n)$.

LEMMA 1. (a) $\widetilde{KO}_q(BZ/2) = \begin{cases} (Z[1/2])/Z & \underline{\text{if}}\ q \equiv 3,7 \bmod 8 \\ Z/2 & \underline{\text{if}}\ q \equiv 1,2 \bmod 8 \\ 0 & \underline{\text{otherwise}} \end{cases}$

(b) $\widetilde{KO}_3(RP^{8n+3}) \simeq Z/2^{2n+3} \oplus Z$

(c) $\widetilde{KO}_3(RP^{8n+4}) \simeq Z/2^{2n+4}$

(d) $\widetilde{KO}_7(RP^{8n+7}) \simeq Z/2^{2n+4} \oplus Z$

(e) $\widetilde{KO}_7(RP^{8n+8}) \simeq Z/2^{2n+5}$.

The proofs are elementary and/or well known, and left to the reader.

LEMMA 2. Let $\iota: RP^{4n+3} \to BZ/2$ be the inclusion. Then $\iota_*[RP^{4n+3}]$ has order 2^{2n+4} if n is even and 2^{2n+5} if n is odd.

PROOF. Consider the homology exact sequence

$$KO_{4n+3}(RP^{4n+3}) \to KO_{4n+3}(RP^{4n+4}) \to KO_{4n+3}(S^{4n+4}) = 0.$$

Then the image of $[RP^{4n+3}]$ must be a generator of $KO_{4n+3}(RP^{4n+4})$ and hence it has the order given above. Since the inclusion $RP^{4n+4} \to BZ/2$ induces a monomorphism in KO homology, the lemma follows.

COROLLARY. The order of $[RP^{4n+3}, \iota]$ in $\Omega_{4n+3}^{Spin}(BZ/2)$ is at least 2^{2n+4} if n is even, and 2^{2n+5} if n is odd.

PROOF. $c(\pi^o)(RP^{4n+3}, \iota) = \iota_*([RP^{4n+3}])$ has the given order. To complete the proof of the theorem, the structure of $\Omega_*^{Spin}(BZ/2)$

is needed. In [3], Anderson, Brown and Peterson computed the group structure, and their methods also give much information about the Ω_*^{Spin} module structure.

From [2], $\Omega_*^{Spin} = \pi_*(MSpin) = (\vee BO\langle 8k\rangle) \vee (\vee BO\langle 8k+2\rangle) \vee K$, where K is a wedge of suspensions of $K(Z/2)$. Then $\tilde{\Omega}_*^{Spin}(BZ/2) = \pi_*(MSpin \wedge BZ/2)$ breaks up into direct summands. In [3] the Adams spectral sequence for each summand was computed. Since the elements in filtration 0 are detected by Stiefel-Whitney numbers, we list those elements in filtration greater than 0. For each summand of the form $BO\langle 8k\rangle \wedge BZ/2$ there are elements corresponding to $h_1^i\, w^j$, $j > 0$, $i = 1,2$ giving elements in dimension $8j + i + 8k$. Also there are elements x_j in filtration 0 with dimension $4j - 1 + 8k$, $j > 0$, such that $h_0^q\, x_j \neq 0$, $0 \leq q \leq 2j + \epsilon$, where $\epsilon = 1$ if j is even, 0 if j is odd. For each summand of the form $BO\langle 8k+2\rangle \wedge BZ/2$ there are elements $h_0^q\, y_j$, y_j with dimension $4j + 3 + 8k$, $0 \leq q \leq 2j + \epsilon - 1$, and classes corresponding to $t_k\, w^j\, h_1^i$ in dimension $8k + 4 + 8j + i$, $i = 1,2$. Let $J = (j_1, \ldots, j_s)$ be a sequence of integers with $j_1 \geq j_2 \geq \ldots \geq j_s > 1$, and $n(J) = j_1 + \ldots + j_s$. Let M_J be the Spin manifolds defined in [2]. For n(J) odd, Let $\hat{M}_J$ be the manifold of dimension $4n(J) + 4$ with $\langle \pi^J\, \pi^1, [\hat{M}_J]\rangle = 1$. From computation of Stiefel-Whitney numbers x_j and y_j can be represented by $M_J \times (RP^{4j-1}, \iota)$. To complete the proof, we must evaluate some charateristic numbers in $KO_*(BZ/2)$.

For n(J) even, since $\langle \pi^J, M_J\rangle = 1$, we have

1) $c(\pi^J)[M_J \times (RP^{4n+3}, \iota)] = \iota_*[RP^{4n+3}] \in KO_{4n+3}(BZ/2)$

2) $c(\pi^J(\pi^1)^{2j}[M_J \times w^j \times (S^1, \iota)] = \iota_*[S^1] \in KO_1(BZ/2)$

3) $c(\pi^J(\pi^1)^{2j})[M_J \times w^j \times S^1 \times (S^1, \iota)] = \iota_*[S^1]^2 \in KO_2(BZ/2)$.

For $n(J)$ odd, equations 2) and 3) hold with $\widetilde{M}_J$ replacing M_J. Also for $n(J)$ odd, we have the following, where $\bar{J} = (J, 1)$:

$$c(\pi^{\bar{J}})[M_J \times (RP^{4n+3}, \iota)] = 2\iota_*[RP^{4n+3}].$$

Thus the KO-characteristic numbers distinguish those classes not distinguished by Stiefel-Whitney numbers, and the theorem is proved.

3. CONSEQUENCES

Let h_* be the homology theory $KO_*(\ \ ) \oplus H_*(\ \ , Z/2)$. A reformulation of the main theorem gives the following corollary,

COROLLARY. <u>The Hurewicz map</u> $\kappa: \tilde{\Omega}_*^{Spin}(BZ/2) \to h_*(MSpin \wedge BZ/2)$ <u>is a monomorphism</u>.

The theorem can also be lifted to the equivariant case.

COROLLARY. $\Omega_*^{Spin}(Z/2, \text{free})$ <u>is detected by equivariant KO and Stiefel-Whitney numbers</u>.

The isomorphism $\zeta: \Omega_n^{Spin} \to \tilde{\Omega}_{n+1}^{Spin}(BZ/2)$ gives a method for detecting the elements of Ω_*^{Pin}. For an element $[M] \in \Omega_n^{Pin}$, simply evaluate the characteristic numbers of $\zeta[M]$. Thus Ω_n^{Pin} is also detected by KO and Stiefel-Whitney characteristic numbers.

REFERENCES

1. D. W. Anderson, E. H. Brown, and F. P. Peterson, Spin Cobordism, Bull. Amer. Math. Soc. 72 (1966), 256-260.

2. D. W. Anderson, E. H. Brown, and F. P. Peterson, The Structure of the Spin Cobordism Ring, Ann. of Math. (2) 86 (1967), 271-298.

3. D. W. Anderson, E. H. Brown, and F. P. Peterson, Pin Cobordism and Related Topics, Comm. Math. Helv. 44 (1969), 462-468.

4. M. F. Atiyah and R. Bott, A Lefschetz Pixed Point Formula for Elliptic Complexes II. Ann. of Math. 88 (1968), 451-491.

5. M. F. Atiyah, R. Bott, and R. Shapiro, Clifford Modules, Topology 3 (1964) Suppl. 1, 3-38.

6. M. F. Atiyah and F. Hirzerbruch, Spin Manifolds and Group Actions, Essays on Topology and Related Topics, Memoires dédiés a Georges de Rham, pp. 18-28. Springer, Berlin and New York, 1970.

UNIVERSITY OF CONNECTICUT
STORRS, CT 06268
U.S.A.

EQUIVARIANT K-THEORY AND CYCLIC SUBGROUPS

STEFAN JACKOWSKI

The works of Atiyah and Segal show that cyclic subgroups are distinguished in equivariant K-theory. The aim of this paper is to prove the theorem which also illustrates this phenomenon:

THEOREM. Let G be a finite group. If the equivariant map $f : X \to Y$ between compact G-CW-complexes induces isomorphisms

$$f^* : K_S^*(Y) \to K_S^*(X)$$

for every cyclic subgroup $S \subset G$, then $f^* : K_G^*(Y) \xrightarrow{\approx} K_G^*(X)$ is also an isomorphism.

This fact is a consequence of the generalized completion theorem in equivariant K-theory. With every family F of subgroups of a compact Lie group G one can associate certain topology, called the F-topology, in the representation ring R(G), and thus also in $K_G^*(X)$, where X is a G-space. On the other hand for every family of subgroups there exists its classifying space EF. The Completion Conjecture states that for every compact G-CW-complex X the projection $X \times EF \to X$ defines an isomorphism

$$\widehat{K_G^*(X)} \xrightarrow{\approx} K_G^*(X \times EF)$$

where $\widehat{\ }$ denotes completion in the F-topology. For the family F consisting of just the trivial subgroup this is the Atiyah-Segal Completion Theorem [3]. In [5] we prove the conjecture in many other cases. In this paper we will consider the completion theorem in case of the finite group and the family of all its cyclic subgroups. It turns out that in this case the topology defined in the representation ring is discrete.

We discuss also the relationship between the orbit structure of a compact G-space X and the properties of the F-topology in $K_G^*(X)$. It is described by the following theorem:

THEOREM. For a compact G-space X the following conditions are equivalent:

a) every cyclic subgroup $S \subset G$ for which the fixed point set X^S is non-empty belongs to the family F,

b) $K_G^*(X)$ is discrete in the F-topology,

c) $K_G^*(X)$ is complete and Hausdorff in the F-topology.

We restrict ourselves here to the case of a finite group, but many of the results are valid also for an arbitrary compact Lie group.

I am grateful to Professor Tammo tom Dieck for helpful conversation.

1. FAMILES OF SUBGROUPS AND EQUIVARIANT COHOMOLOGY

Let G be a finite group and let F be a family of subgroups of G (see [4; Def. 1]). For a given family F, a G-space X is called F-free iff all isotropy subgroups occurring on X belong to F.

Throughout the paper we shall assume that a G-space X is a G-CW-complex (in the sense of Matumoto [7]). The final object in the G-homotopy category of F-free G-CW-complexes is called the classifying space for the family F (see [4; Def. 3]). For every family F the classifying space exists and can be obtained as an infinite join of orbits ([4; Satz 1]) with compactly generated topology. We will denote it by EF.

In the sequel we will need the following properties of classifying spaces.

Remark that for a subgroup $H \subset G$ and a family F the set of subgroups $F \cap H := \{K \leq H | K \in F\}$ is a family of subgroups of H.

1.1 PROPOSITION. *If* $H \leq G$ *then for arbitrary family of subgroups* F *of a group* G *there exists an* H-*homotopy equivalence of classifying spaces*

$$EF = E(F \cap H).$$

1.2 PROPOSITION. *If* F_1, F_2 *are families of subgroups of* G *then there exists a G-homotopy equivalence*

$$E(F_1 \cap F_2) = EF_1 \times EF_2$$

where the topology on the cartesian product is the smallest compactly generated topology containing Tychonoff topology.

For every equivariant cohomology theory h_G^* and every pair of families of subgroups $F_1 \supset F_2$ of the group G tom Dieck [4] defined a new theory $h_G^*[F_1, F_2]$. Theories $h_G^*[F_1, F_2]$ "detect" at most those orbits whose isotropy subgroups belong to $F_1 \setminus F_2$. The properties of theories $h_G^*[All, F]$ were investigated by tom Dieck in [4]. Here we will study theories of the form $h_G^*[F]$. We assume that the theory h_G^* is defined on the category of all G-CW-complexes and is additive, so Milnor's lemma holds. We will need also the existence of the natural isomorphisms

$$h_G^*(G \times_H X) \cong h_H^*(X)$$

for every subgroup $H \leq G$ and H-space X.

In order to describe the properties of the theory $h_G^*[F]$ it is convenient to introduce the following definition:

1.3 DEFINITION. A G-map $F : X \to Y$ is called an F-cohomology equivalence iff it induces isomorphisms

$$f^* : h_H^*(Y) \to h_H^*(X)$$

for every subgroup $H \leq F$.

1.4 PROPOSITION. *Let* E *be an F-free space and let* $f : X \to Y$ *be an F-cohomology equivalence between compact G-spaces. Then the induced homormophism* $(\mathrm{id} \times f)^* : h_G^*(E \times Y) \to h_G^*(E \times X)$ *is an isomorphism.*

PROOF. Let us assume that E is a compact G-space and consider the diagram over E/G:

$$\begin{array}{ccccc}
E \times X & & \xrightarrow{\mathrm{id} \times f} & & E \times Y \\
p \downarrow & & & & \downarrow p' \\
E & \xrightarrow{\pi} & E/G & \xleftarrow{\pi} & E
\end{array}$$

The map $\mathrm{id} \times f$ induces the homomorphism of Segal's spectral sequences [9] for the maps πp and $\pi p'$ and it is easy to verify that this homormorphism is an isomorphism on the E_2-terms. To prove the proposition for a non-compact space E we apply Milnor's lemma.

This is a theorem of Vietoris-Begle-type in the equivatiant case. See also Kosniowski [6; Theorem 2.14].

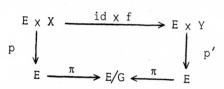

1.5 THEOREM. The natural transformation $h_G^* \to h_G^*[F]$ has the following properties:

a) If X is an F-free space then $h_G^*(X) \approx h_G^*[F](X)$,

b) If $f : X \to Y$ is an F-cohomology equivalence between compact G-spaces then $f^* : h_G^*[F](Y) \to h_G^*[F](X)$ is an isomorphism.

PROOF. The first property follows easily from tom Dieck's results [4; Satz 6], while the second one is a direct consequence of the Proposition 1.6.

It can be proved that these two properties give a functorial characterisation of the theory $h_G^*[F]$ on the category of G-CW-complexes.

2. TOPOLOGIES IN THE REPRESENTATION RING

We will describe how with every family of subgroups of the group G one can associate a certain topology in the representation ring R(G).

For a given subgroup $H \leq G$, I(H) will denote the kernel of the restriction homomorphism $R(G) \to R(H)$. The set of ideals

$$I(F) := \{I(H) \cdot \ldots \cdot I(H_n) | H_i \in F\}$$

defines the basis of neighbourhoods of zero in the representation ring. The corresponding topology will be called the F-topology in R(G). Because we have assumed that the group G is finite this topology is defined by the ideal $I(H_1) \cdot \ldots \cdot I(H_k)$ where $F = \{H_1, \ldots, H_k\}$.

For the given family F we will denote by $I'(F)$ the kernel of the restriction homomorphism

$$R(G) \to \prod_{H \in F} R(H)$$

2.1 PROPOSITION. *The* F-*topology in* R(G) *coincides with* $I'(F)$-*adic topology.*

PROOF. Because obviously $I'(F) = I(H_1) \cap \ldots \cap I(H_k)$ and the ring R(G) is noetherian then the proposition follows from the fact that the radicals $r(I'(F)) = r(I(H_1) \cdot \ldots \cdot I(H_k))$ are equal.

Observe that the last proposition implies that for every family F the F-topology coincides with the F^c-topology, where F^c is the family of all cyclic subgroups contained in F. Moreover if the family F contains all cyclic subgroups of G then the F-topology in R(G) is discrete. In view of the next proposition this gives a complete description of discrete F-topologies in R(G).

2.2 PROPOSITION. *The following conditions are equivalent*:

a) *the family* F *contains all cyclic subgroups*,

b) *the F-topology in* R(G) *is discrete*,

c) *the F-topology in* R(G) *is complete and Hausdorff*.

PROOF. (a) ⇒ (b) ⇒ (c) is clear.

(c) ⇒ (a) if the ring R(G) is complete and Hausdorff in the F-topology then the ideal $I'(F)$ is contained in the Jacobson radical of R(G) (cf. [2; prop. 10.15]). The ring R(G) is finitely generated $\mathbb{Z}$-module thus its Jacobson radical coincides with its nilradical (cf. [2; ex. 5.24]) - but the last one is zero because there are no nilpotent elements in R(G). Therefore we know that

$I'(F) = 0$ and the restriction homomorphism $R(G) \to \Pi R(H)$ is a monomorphism, which implies that all cyclic subgroups of G are contained in F.

Let H be a subgroup of G and let F be a family of subgroups of G. We will now compare two topologies in R(H): the $F \cap H$-topology and the F-topology which is determined by the F-topology in R(G) and the restriction homomorphism $R(G) \to R(H)$.

2.3 PROPOSITION. _The F-topology on_ R(H), _when it is regarded as an_ R(G)-_module, coincides with its_ $F \cap H$-_topology_.

PROOF. From the description of prime ideals in the representation ring given by Segal [8] it follows that for the inclusion $i : H \hookrightarrow G$ and a prime ideal $p \in \text{Spec } R(H)$ whose support is $S_p \le H$, the same group S_p is a support for the ideal $i_*(p) \in \text{Spec } R(G)$. This fact shows that the radicals of the ideals $I'(F \cap H)$ and $i^*(I'(F)) \cdot R(H)$ coincide.

Using Atiyah's methods [1] one can describe the kernel of the completion homomorphism in F-topology: $R(G) \to R(G)^{\wedge}$. We will say that an element $g \in G$ belongs to F if and only if the cyclic subgroup generated by g is contained in F.

2.4 THEOREM. _The kernel of the completion homomorphism in the_ F-_topology_ $R(G) \to R(G)^{\wedge}$ _consists of characters_ $\chi \in R(G)$ _vanishing on those elements_ $g \in G$, _for which there exists a prime number_ p _such that_ $g^{p^r} \in F$ _for some integer_ r.

3. COMPLETION OF EQUIVARIANT K-THEORY

Let F be a family of subgroups of the group G. We have associated with that family the F-topology in the ring R(G), which will be now regarded as coefficients of equivariant K-theory. For an arbitrary G-space X the projection $X \to pt$ and the F-topology in R(G) determine F-topology in the ring $K_G^*(X)$. Homomorphisms induced by equivariant maps are continuous in the F-topology. Let us also remark that the canonical isomorphism $K_G^*(G/H) = K_H^*(pt) = R(H)$ allows us to regard the ideals I(H), used in the definition of the F-topology, as kernels of the homomorphisms $K_G^*(pt) \to K_G^*(G/H)$ induced by the projections $G/H \to pt$.

The following proposition gives the basic connection between the orbit structure of a G-space X and the F-topology in $K_G^*(X)$.

3.1 PROPOSITION. _If the compact_ G-_space_ X _is_ F-_free then the_ F-_topology in_ $K_G^*(X)$ _is discrete_.

PROOF. The space X is compact thus $X = U_{x_1} \cup \ldots \cup X_{x_k}$, where $U_{x_i} \supset Gx_i$ are the tubes around the orbits Gx_i. From the definition the images of the ideals $I(G_{x_i})$ under the homomorphisms $K_G^*(pt) \to K_G^*(U_{x_i})$ are zero. Thus the image of $I(G_{x_1})\cdot\ldots\cdot I(G_{x_k})$ by the homormorphism $K_G^*(pt) \to K_G^*(X)$ also is zero.

Let EF be a classifying space for the family F and let $\{Y\}$ denote the collection of its G-invariant compact subsets. Following Atiyah and Segal [3] we will formulate the next proposition using pro-rings.

3.2. PROPOSITION. _For an arbitrary compact_ G-_space_ X _the projection_ $X \times EF \to X$ _defines the homomorphism of pro-rings_

$$\alpha : \{K_G^*(X)/I\cdot K_G^*(X) | I \in I(F)\} \to \{K_G^*(X \times Y) | Y \subset EF\}$$

PROOF. For any $Y \in \{Y\}$ let $\alpha_Y : X \times Y \to X$ be a projection. From the previous proposition we know that $K_G^*(X \times Y)$ is discrete in the F-topology, so there exists an ideal $I_Y \in I(F)$ such that

$$\alpha_Y^* : K_G^*(X)/I_Y K_G^*(X) \to K_G^*(X \times Y)$$

The family of homomorphisms $\{\alpha_Y^*\}$ gives the desired homomorphism of pro-rings.

We will now formulate the Completion Conjecture, whose discussion can be found in [5].

3.3 CONJECTURE. The homomorphism α induces the natural isomorphism of cohomology theories on the category of compact G-CW-complexes

$$\widehat{K_G^*(X)} \overset{\approx}{\to} K_G^*[F](X)$$

where $\widehat{\ }$ denotes the completion in the F-toplogy.

In this paper we restrict ourselves to the case when F is a family of all cyclic subgroups. To prove the completion theorem in this case it will be necessary to consider families determined by appropriate representations.

Let V be a complex representation of the group G. We will denote F_V the family of subgroups of F defined by the representatior V (see [4; §4]). We will seek for description of the F_V-topology in R(G) in terms of the Euler class e(V) of the representation V (see [4; §4]). Consider the family of principal ideals $e(V)^n K_G^*$ and the suitable topology, which will be called e(V)-topology.

3.4 LEMMA. <u>The</u> e(V)-<u>topology coincides with the</u> F_V-<u>topolo</u>

PROOF. For every $H \in F_V$ there exists an equivariant map $G/H \to V \setminus \{0\}$. This implies that $e(V) \in I(H)$ and therefore the F_V-topology is weaker than the $e(V)$-topology.

Conversely: we will consider the Gysin sequence for the representation V:

$$\dots \to K_G^*(pt) \xrightarrow{e(V)} K_G^*(pt) \xrightarrow{\pi^*} K_G^*(s(V)) \to \dots$$

where s(V) denotes the sphere in V with respect to some equivariant metric. The sphere s(V) is compact thus $s(V) = U_{x_1} \cup \dots \cup U_{x_k}$ and $U_{x_i} \supset Gx_i$ are tubes around the orbits. Hence $I(G_{x_1}) \cdot \dots \cdot I(G_{x_k}) \subset \ker \pi^* = e(V) \cdot K_G^*$ which finishes the proof.

3.5 PROPOSITION. _If_ X _is compact G-space such that_ $K_G^*(X)$ _is a finite generated_ R(G)-_module then the homomorphism_

$$\bar{\alpha} : \{K_G^*(X)/I \cdot K_G^*(X) \mid I \in I(F_V)\} \to \{K_G^*(X \times Y) \mid Y \subset EF_V\}$$

is the isomorphism of pro-rings.

PROOF. The classifying space EF_V is the sphere in the infinite dimensional representation: $EF_V = \lim s(V \oplus .. \oplus V)$. Now we apply the Gysin sequence and using Lemma 3.4 we proceed as in the proof of Lemma 3.1 of Atiyah and Segal [3].

3.6 COROLLARY. _If the assumptions of the Proposition 3.5 are fulfilled then_ $\bar{\alpha}$ _defines an isomorphism_

$$\widehat{K_G^*(X)} \overset{\approx}{\to} K_G^*[F_V](X)$$

where $\widehat{\ }$ _denotes the completion in the_ F_V_-topology_.

PROOF. cf. Atiyah and Segal [3; proposition 4.2].

The above corollary can be regarded as a statement dual to tom Dieck's result ([4; Satz 5]) dscribing in homotopy terms the localisation of the homology theory with respect to Euler classes.

4. THE FAMILY OF CYCLIC SUBGROUPS

We will now prove the completion theorem for the family All^c of all cyclic subgroups of the group G. We have observed that the All^c-topology in the representation ring is discrete so we have to prove that the projection $EAll^c \to pt$ induces an isomorphism of cohomology theories

$$K_G^*(\cdot) \to K_G^*[All^c](\cdot)$$

We will denote by All_p the family of all proper subgroups of a given group.

4.1 LEMMA. <u>If the group</u> G <u>is non-cyclic, then the projection</u> $EAll_p \to pt$ <u>induces the isomorphism of cohomology theories</u>

$$K_G^*(\cdot) \to K_G^*[All_p](\cdot)$$

<u>on the category of all</u> G-CW-<u>complexes</u>.

PROOF. For an arbitrary group G the family All_p is determined by the representation which is the sum of all non-trivial irreducible representations of G. If G is non-cyclic, then $All_p \supset All^c$ and (cf. Proposition 2.2) the All_p-topology is discrete. From the Corollary 3.6 we obtain the natural isomorphism

$$K_G^*(X) \to K_G^*[All_p](X)$$

on the category of finite G-CW-complexes. Notice, that the theory $K_G^*[All_p]$ is additive and thus the above isomorphism holds also for infinite G-CW-complexes.

We recall that for finite group G there exists a chain of families of subgroups $\{e\} = F_o \subset F_1 \subset \ldots \subset F_n = \text{All}$ such that $F_i \setminus F_{i-1} = (H_i)$ consists of the subgroups conjugate to a fixed subgroup H_i. Such families are called adjacent.

4.2 THEOREM. <u>The projection</u> $EAll^c \to pt$ <u>induces the isomorphism of cohomology theories</u>

$$K_G^*(\cdot) \to K_G^*[All^c](\cdot)$$

<u>on the category of all</u> G-CW-<u>complexes</u>.

PROOF. We will use the induction method introduced by Stong and tom Dieck.

Let $\{e\} = F_o \subset F_1 \subset \ldots \subset F_n = \text{All}$ be a sequence of adjacent families in the group G. For F_o we have the isomorphism

$$T_o : K_G^*[\{e\}] \to (K_G^*[All^c])[\{e\}] = K_G^*[All^c \cap \{e\}].$$

Suppose that the homomorphism $T_{i-1} : K_G^*[F_{i-1}] \to K_G^*[All^c \cap F_{i-1}]$ is an isomorphism. We will prove that T_i is an isomorphism. To prove this assertation it is sufficient to prove that the relative groups

$$K_G^*[F_i, F_{i-1}] \xrightarrow{\approx} K_G^*[F_i^c, F_{i-1}^c]$$

where $F_k^c := All^c \cap F_k$, are isomorphic.

We will distinguish two cases:

i) the subgroup $(H) = F_i \setminus F_{i-1}$ is not cyclic. Then $K_G^*[F_i^c, F_{i-1}^c](X)$ is zero because $F_i^c = F_{i-1}^c$. We must show that $K_G^*[F_i, F_{i-1}](X) = 0$. T. tom Dieck's result [4; Satz 4] provides a useful description of the relative groups:

$$K_G^*[F_i, F_{i-1}](X) = K_G^*(E \times (CEF_{i-1}, EF_{i-1}) \times X)$$

where $E := G \times_{NH} E(NH/H)$ and CEF_{i-1} denotes the cone over EF_{i-1}. On the other hand we have:

$$K_G^*(G/H \times (CEF_{i-1}, EF_{i-1}) \times X) = K_G^*(G \times_H ((CEF_{i-1}, EF_{i-1}) \times X))$$

$$= K_H^*((CEF_{i-1}, EF_{i-1}) \times X) = K_H^*[All, All_p](X)$$

because from the definition of adjacent families it follows that $F_{i-1} \cap H = All_p$. From Lemma 4.1, and the long exact sequence we deduce that $K_H^*[All, All_p](X) = 0$. The only orbit type occurring on E is G/H, so for arbitrary finite subcomplex $Z \subset E$ we have

$$K_G^*(Z \times (CEF_{i-1}, EF_{i-1}) \times X) = 0$$

This equality and Milnor's lemma imply that $K_G^*[F_i, F_{i-1}](X) =$

ii) the subgroup $(H) = F_i \setminus F_{i-1}$ is cyclic. We argue similarly as is (i):

$$K_G^*[F_i, F_{i-1}](X) = K_G^*(E \times (CEF_{i-1}, EF_{i-1}) \times X)$$

$$K_G^*[F_i^c, F_{i-1}^c](X) = K_G^*(E \times (CEF_{i-1}^c, EF_{i-1}^c) \times X)$$

From the induction hypothesis we have that the inclusion $F_{i-1}^c \hookrightarrow F_{i-1}$ induces an isomorphism in K_G^*. Thus we have

$$K_G^*[F_i, F_{i-1}](X) = K_G^*[F_i^c, F_{i-1}^c](X)$$

which finishes proof of the theorem.

The corollary which we will derive from the completion theorem for the family of all cyclic subgroups is another illustration of the fact that cyclic subgroups are distinguished in equivariant K-theory.

4.3 COROLLARY. <u>If the map</u> $f: X \to Y$ <u>between compact</u> G-CW-<u>complexes is an</u> All^c-<u>cohomological equivalence in</u> K_G-<u>theory, then</u>

it induces an isomorphism $f: K_G^*(Y) \xrightarrow{\approx} K_G^*(X)$.

PROOF. Follows from Theorem 4.2 and 1.5.

The analogous corollary for K_G^*-theory tensored with rationals follows from the Atiyah's conjecture proved by R. Rubinsztein (unpublished):

4.4 THEOREM. *For a compact Lie group* G *and an arbitrary finite* G-CW-*complex* X *the restriction homomorphism*

$$K_G^*(X) \otimes Q \to \prod_{S \in All^c} K_S^*(X) \otimes Q$$

is a monomorphism.

We can obtain Corollary 4.3 for $K_G^*(\cdot) \otimes Q$ from this Theorem replacing X by a cone of a map f. However, Theorem 4.4 is not true without tensoring with Q even for the trivial G-spaces. The Corollary 4.3 can be regarded as the strongest version of Atiyah's conjecture without tensoring with Q.

5. COMPLETENESS AND ORBIT STRUCTURE

We will discuss what information about the orbit structure of a compact G-space X can be obtained from the F-topology in $K_G^*(X)$.

5.1 THEOREM. *Let* F *be a family of subgroups of the finite group* G. *For a compact* G-*space* X *the following conditions are equivalent:*

a) *every cyclic subgroup* $S \subset G$ *for which the fixed point set* X^S *is non-empty, belongs to the family* F,

b) $K_G^*(X)$ *is discrete in the* F-*topology,*

c) $K_G^*(X)$ *is complete and Hausdorff in the* F-*topology.*

PROOF. (a) ⇒ (b). Let us denote by F_X the smallest family containing all isotropy subgroups occurring on X. Then from the Proposition 3.1 we deduce that $K_G^*(X)$ is discrete in the F_X-topology. The assumption about the orbit structure of X means that $F_X^c \subset F$. The F_X^c-topology coincides with the F_X-topology (cf. section 2) so the ring $K_G^*(X)$ is discrete in the F_X^c-topology and hence in the F-topology.

(b) ⇒ (c) is obvious.

(c) ⇒ (a) Let S be a cyclic subgroup such that $X^S \neq \phi$. Consider the homomorphism $K_G^*(X) \to R(S)$ induced by an equivariant map $G/S \to X$. This homomorphism imposes on R(S) a structure of a finitely generated $K_G^*(X)$-module. Therefore if $K_G^*(X)$ is complete and Hausdorff in the F-topology the same holds for R(S). From Proposition 2.3 it follows that the F-topology on R(S) coincides with the $F \cap S$-topology so we deduce (cf. Proposition 2.2) that $S \in F$.

REFERENCES

1. M. F. Atiyah, Characters and cohomology of finite groups. Publ. Math. IHES 9 (1961), 23-64.

2. M. F. Atiyah, I. Macdonald, Introduction to commutative algebra. Addison-Wesley Reading, Mass., 1969.

3. M. F. Atiyah, G. B. Segal, Equivariant K-theory and completi J. Diff. Geom. 3 (1969), 1-18.

4. T. tom Dieck, Orbittypen und aequivariante Homologie. Arch. Math. XXIII (1972), 307-317.

5. S. Jackowski, Families of subgroups and completion. To appear.

6. C. Kosniowski, Equivariant Cohomology and Stable Cohomology. Math. Ann. 210 (1974) 83-104.

7. T. Matumoto, On G-CW-complexes and theorem of J.H.C. Whitehead. J. Fac. Sci. Univ. Tokyo Sect. 1. 18 (1971), 363-374.

8. G. B. Segal, The representation ring for the compact Lie group. Publ. Math. IHES 34 (1968), 113-128

9. G. B. Segal, Equivariant K-theory. Publ. Math. IHES 34 (1968), 129-151.

WARSAW UNIVERSITY,
PKin IXp,
00-901 WARSZAWA,
POLAND.

$\mathbb{Z}/p$ MANIFOLDS WITH LOW DIMENSIONAL FIXED POINT SET

CZES KOSNIOWSKI

1. INTRODUCTION

This paper contains a description of the generators of $\mathfrak{U}_*^{\mathbb{Z}/p}(r)$, where $\mathfrak{U}_*^{\mathbb{Z}/p}(r)$ is the subring of the unitary $\mathbb{Z}/p$ bordism ring $\mathfrak{U}_*^{\mathbb{Z}/p}$ generated by $\mathbb{Z}/p$ manifolds with fixed point set of dimension less than 2r. Only the case p is prime and $r \leq p-1$ will be considered throughout this paper. A corollary of the main result obtained verifies a conjecture of the author in [6], (for the case p prime and $r \leq p-1$).

The result for $\mathfrak{U}_*^{\mathbb{Z}/p}(0)$ is well known since this ring consists of free $\mathbb{Z}/p$ manifolds. An "algebraic" result for r=1 (the case of isolated fixed points) was given in [4]. Furthermore, for the special case of p=2, 3 explicit geometric generators of $\mathfrak{U}_*^{\mathbb{Z}/p}(1)$ were given in [4] as Theorems 2.2, 2.6. If p=3 or 5 then [3] describes geometric generators of $\Omega_*^{\mathbb{Z}/p}(1)$ - the oriented analogue of $\mathfrak{U}_*^{\mathbb{Z}/p}(1)$.

The generators of $\mathfrak{U}_*^{\mathbb{Z}/p}(r)$ are some "obvious" generators - $\mathbb{Z}/p$, Riemann surfaces and complex projective spaces (with $\mathbb{Z}/p$ action) - together with some $\mathbb{Z}/p$ manifolds obtained from these via a "λ construction". Suppose M is a $\mathbb{Z}/p$ manifold with an action Φ of the torus $(S^1)^\ell$ which commutes with the $\mathbb{Z}/p$ action (i.e. an action of $\mathbb{Z}/p \times (S^1)^\ell$ on M). The λ construction produces a $\mathbb{Z}/p$

manifold $\lambda_\ell(M, \Phi)$ of dimension $2\ell + \dim M$. Under suitable restrictions on M, ℓ and Φ the $\mathbb{Z}/p$ manifold $\lambda_\ell(M, \Phi)$ has no higher dimensional fixed point set component than M has. Recall that in [5] the "Γ construction" was used to obtain generators of $\mathfrak{N}_*^{\mathbb{Z}/p}$. If the $\mathbb{Z}/p$ action on M extends to an S^1 action and if $\Phi|S_i^1$ (where $(S^1)^\ell = \prod_{i=1}^{\ell} S_i^1$) coincides with this action for $i = 1,2,\ldots,\ell$ then $\lambda_\ell(M, \Phi) = \Gamma^\ell(M)$. For the purpose of this paper the Γ construction is of little value since $\Gamma^\ell(M)$ has a codimension 2 fixed point set component, namely $\Gamma^{\ell-1}(M)$. Section 2 contains further details.

If p=2 then the condition $r \leq p-1$ means that r = 0 or 1. The ring $\mathfrak{U}_*^{\mathbb{Z}/p}(0)$ is well understood and $\mathfrak{U}_*^{\mathbb{Z}/p}(1)$ is described adequately in [4]. Therefore, throughout this paper p is an odd prime.

The proof of the main result, i.e. that a certain list of $\mathbb{Z}/p$ manifolds generates $\mathfrak{U}_*^{\mathbb{Z}/p}(r)$, is roughly as follows. A natural map $\nu : \mathfrak{U}_*^{\mathbb{Z}/p}(r) \to \mathbb{Z}[x_1, x_2, \ldots, x_{rp-1}]$ is defined with kernel precisely the set of free $\mathbb{Z}/p$ manifolds. Thus $\mathfrak{N}_*^{\mathbb{Z}/p}(r)/\mathfrak{N}_*^{\mathbb{Z}/p}(0)$ is isomorphic to a subring of the polynomial ring $\mathbb{Z}[x_1, x_2,\ldots,x_{rp-1}]$. Rationally the rings $\mathfrak{U}_*^{\mathbb{Z}/p}(r)/\mathfrak{N}_*^{\mathbb{Z}/p}(0) \otimes Q$ and $Q[x_1, x_2, \ldots, x_{rp-1}]$ are isomorphic. The polynomial ring $\mathbb{Z}[x_1, x_2, \ldots, x_{rp-1}]$ is graded by saying that $\deg x_j = [(j + p-1)/p]$. The homogeneous elements of degree n form a $\mathbb{Z}$ module $\mathcal{M}(n, r)$ which contains the image under ν of $\mathfrak{N}_{2n}^{\mathbb{Z}/p}(r)$. Let $\mathcal{J}(n, r)$ denote this image in $\mathcal{M}(n, r)$. The "list of generators" give an upper bound for $|\mathcal{M}(n, r)/\mathcal{J}(n, r)|$ (which is finite by the remark on rationarity above). Various integrality conditions give a lower bound for $|\mathcal{M}(n, r)/\mathcal{J}(n, r)|$. These bounds coincide and hence show that the list of generators do indeed generate $\mathfrak{U}_*^{\mathbb{Z}/p}(r)$.

These last steps in the proof may remind some readers of the last steps in the proof of the main theorem of [4]. The value of $|\mathcal{M}(n, r)/\mathcal{I}(n, r)|$ is $\Pi\ p^{q(I)}$ where the product is over unordered sequences of integers $I = (i(1), i(2), \ldots, i(k))$, $n \geq k \geq 0$, satisfying

(i) $1 \leq i(j) \leq rp-1$ for $j = 1, 2, \ldots, k$,

(ii) $i(j) \neq p-1$ for $j = 1, 2, \ldots, k$ and

(iii) $\Sigma_{j=1}^{k}[(i(j) + p-1)/p] \leq n$

while $q(I) = [(n + p-2 - \Sigma i(j))/(p-1)]$.

The next section, section 2, gives details of the generators. Sections 3 and 4 contain the essential ingredients to give the upper bound for $|\mathcal{M}(n, r)/\mathcal{I}(n, r)|$ while section 5 contains the ingredients to give the lower bound. The final details are collected in section 6. Section 7 describes some very interesting consequences.

2. THE GENERATORS

This section contains a description of the generators of $\mathfrak{U}_*^{\mathbb{Z}/p}(r)$. It is well known that $\mathfrak{U}_*^{\mathbb{Z}/p}(0)$, (the subring of $\mathfrak{U}_*^{\mathbb{Z}/p}$ consisting of free $\mathbb{Z}/p$ manifolds), is precisely the subring $\mathbb{Z}/p \cdot \mathfrak{U}_*$ of $\mathfrak{U}_*^{\mathbb{Z}/p}$. Elements of $\mathbb{Z}/p \cdot \mathfrak{U}_*$ are of the form $\mathbb{Z}/p \times M$ with $M \in \mathfrak{U}_*$. The action of $\mathbb{Z}/p$ on M being trivial while on $\mathbb{Z}/p$ it is the natural free action. If we write $\mathfrak{U}_*$ as $\mathbb{Z}[M_1, M_2, \ldots]$ and if M_I means $\prod_j M_{i(j)}$ where $I = (i(1), i(2), \ldots, i(k))$ is some unordered k-tuple then $\mathfrak{U}_*^{\mathbb{Z}/p}(0)$ is the subring of $\mathfrak{U}_*^{\mathbb{Z}/p}$ generated by the following set of $\mathbb{Z}/p$ manifolds

$$\{\mathbb{Z}/p \times M_I;\ \text{I is an unordered k tuple},\ k \geq 0\}.$$

THEOREM 2.1. $\mathfrak{U}_*^{\mathbb{Z}/p}(1)$ _is the subring of_ $\mathfrak{U}_*^{\mathbb{Z}/p}$ _generated by the following_ $\mathbb{Z}/p$ _manifolds_

$\{\text{generators of } \mathfrak{U}_*^{\mathbb{Z}/p}(0)\}$

$\cup\{\text{point}\}$

$\cup\{R_j;\ (p+1)/2 \leq j \leq p-1\}$

$\cup\{\lambda_\ell CP^1(p-j);\ 1 \leq j \leq (p-1)/2,\ 0 \leq \ell \leq p-1-2j\}$

$\cup\{\lambda_\ell CP^2(j,\ 1);\ (p+1)/2 \leq j \leq p-1,\ 0 \leq \ell \leq 2j-p-1\}.$

THEOREM 2.r. $(2 \leq r \leq p-1)$. $\mathfrak{U}_*^{\mathbb{Z}/p}(r)$ _is the subring of_ $\mathfrak{U}_*^{\mathbb{Z}/p}$ _generated by the following_ $\mathbb{Z}/p$ _manifolds_

$\{\text{generators of } \mathfrak{U}_*^{\mathbb{Z}/p}(r-1)\}$

$\cup\{M_{r-1}\}$

$\cup\{\lambda_\ell CP^{r-1}(0,\ 0,\ldots,0);\ 0 < \ell \leq (r-1)(p-1)\}$

$\cup\{\lambda_\ell CP^r(0,\ 0,\ldots,0,\ j);\ 1 \leq j \leq p-1,\ 0 \leq \ell \leq (r-1)(p-1) + j-1\}$

In other words, any element of $\mathfrak{U}_*^{\mathbb{Z}/p}(r)$ may be written (not uniquely) as a polynomial in the list of manifolds given in Theorem 2.r, $(r = 1,\ 2,\ldots,p-1)$.

The notation used in the above theorems is explained in the remainder of this section.

(i) point - the single point with trivial $\mathbb{Z}/p$ action.

(ii) $R_j((p+1)/2 \leq j \leq p-1)$ - this is the Riemann surface of genus $(r(j) - 1)(p-1)/2$ associated to the complex function $u = (z^p-1)^{1/r(j)}$ where $r(j)$ is the unique integer satisfying $j\,r(j) = 1 \bmod p$ and $0 < r(j) < p$. The action $\mathbb{Z}/p$ on R_j is induced by $z \to \xi^{-1}z$ where $\xi = \exp(2\pi i/p)$. (An alternative description of R_j may be found by looking in section 5 of [5] and making some appropriate sign changes.) Henceforth we shall use the symbol r_j for $r(j)$.

(iii) $CP^k(J)$, where k is a natural number and $J = (j(1), j(2),\ldots,j(k))$ is a k-tuple of integers, is the complex projective k space CP^k with $\mathbb{Z}/p$ action given by

$$[z_0; z_1; z_2;\ldots;z_k] \to [z_0; z_1\ \xi^{j(1)}; z_2\ \xi^{j(2)};\ldots; z_k\ \xi^{j(k)}]$$

Finally, we need to explain $\lambda_\ell CP^k(J)$. More generally we shall define $\lambda_\ell(M, \Phi)$ where M is a $\mathbb{Z}/p$ manifold and Φ is an action of the torus $(S^1)^\ell$ on M commuting with the $\mathbb{Z}/p$ action on M. As a manifold $\lambda_\ell(M, \Phi)$ is the quotient of $M \times (S^3)^\ell$ by a free $(S^1)^\ell$ action. If we write $(S^d)^\ell$ as $\Pi_{j=1}^\ell S_j^d$ where $S_j^d = S^d$ and $d = 1$ or 3, then an element of $M \times (S^3)^\ell$ may be written as $m \times \Pi(u_j, v_j)$ with $m \in M$ and $(u_j, v_j) \in S_j^3 \subset C^2$. An element $\Pi t_j \in \Pi S_j^1$ acts on $M \times (S^3)^\ell$ by

$$m \times \Pi(u_j, v_j) \to \Phi(\Pi t_j, m) \times \Pi(t_j\ u_j, t_j^{-1}\ v_j).$$

This action of $(S^1)^\ell$ on $M \times (S^3)^\ell$ is free and so the quotient $(M \times (S^3)^\ell)/(S^1)^\ell$ is a manifold. The group $\mathbb{Z}/p$ acts on $M \times S^3$ as follows

$$m \times \Pi(u_j, v_j) \to \xi m \times (u_j, \xi^{-1}\ v_j)$$

where $\xi \in \mathbb{Z}/p \subset S^1$ and ξm denotes the action of ξ on $m \in M$. This induces an action on the quotient and the resulting $\mathbb{Z}/p$ manifold is denoted by $\lambda_\ell(M, \Phi)$.

REMARK. If $\ell = 1$ and Φ is an extension of the $\mathbb{Z}/p$ action on M then $\lambda_1(M, \Phi) = \Gamma(M)$ as defined in [5] etc. If $\ell = 1$ and Φ is not an extension of the $\mathbb{Z}/p$ action on M then $\lambda_1(M, \Phi) = \Gamma(M) - \Gamma(M')$ up to $\mathbb{Z}/p$ bordism where M' is a manifold diffeomorphic to M but with a (possibly) different $\mathbb{Z}/p$ action.

So, to define $\lambda_\ell CP^k(J)$, we need to define some torus actions on CP^k. Suppose $L = (\ell(1), \ell(2),\ldots,\ell(k))$ is a k-tuple of non-negative integers such that $\ell(1) + \ell(2) + \ldots + \ell(k) = \ell$. Now write $(S^1)^\ell$ as $\prod_{i=1}^{k} \prod_{j=1}^{\ell(i)} S^1_{i,j}$ where $S^1_{i,j} = S^1$ and define Φ_L to be the action of $(S^1)^\ell$ on CP^k given by

$$[z_o;z_1;z_2;\ldots;z_k] \rightarrow [z_o;z_1 \prod_{j=1}^{\ell(1)} t_{1,j};\ z_2 \prod_{j=1}^{\ell(2)} t_{2,j};\ldots;z_k \prod_{j=1}^{\ell(k)} t_{k,j}]$$

where $t_{i,j} \in S^1_{i,j}$.

Denote $\lambda_\ell(CP^k(J), \Phi_L)$ by $\Lambda_L CP^k(J)$. Furthermore, if $L = (\ell(1), \ell(2),\ldots,\ell(k))$ is given by

$$\ell(j) = \begin{cases} p-1 & \text{if } 1 \le j \le [\ell/(p-1)] \\ \ell - (p-1)[\ell/(p-1)] & \text{if } j = [\ell/(p-1)] + 1 \\ 0 & \text{otherwise} \end{cases}$$

then denote $\lambda_\ell(CP^k(J), \Phi_L) = \Lambda_L CP^k(J)$ by $\lambda_\ell CP^k(J)$.

3. FIXED POINT SETS AND λ.

In this section we shall describe a natural map ν from $\mathfrak{U}_*^{\mathbb{Z}/p}$ to the polynomial ring $\mathbb{Z}[x_1, x_2,\ldots]$. This map is defined in terms of the equivariant normal bundle of the fixed point set. Restricted to $\mathfrak{U}_*^{\mathbb{Z}/p}(r)$, this map ν factors through the polynomial ring $\mathbb{Z}[x_1, x_2,\ldots, x_{rp-1}]$. Order the monomials in this ring by ordering the variable $\{x_i\}$ so that $x_1 < x_2 < \ldots < x_{rp-1}$, and let lots denote lower order terms. By using the "generators of $\mathfrak{U}_*^{\mathbb{Z}/p}(r)$" we shall prove the following result in this section.

THEOREM 3.1. *For each pair of integers* (i, ℓ) *satisfying* $p \le i \le rp-1$ *and* $0 \le \ell \le i - [(i+p-1)/p]$ *there is a* $\mathbb{Z}/p$ *manifold, obtainable from the "generators" of* $\mathfrak{U}_*^{\mathbb{Z}/p}(r)$, *whose image under* ν *is* $(x_{p-1})^{\ell} x_i$ + lots.

A corresponding result for the case $1 \le i \le p-2$ is proved in the next section.

Let M be a unitary $\mathbb{Z}/p$ manifold. Let F be a component of the fixed point set. The normal bundle $\nu(F, M)$ of F in M has a natural complex structure and splits into a direct sum

$$\nu(F, M) = \Sigma_{j=1}^{p-1} \nu_j(F)$$

of (complex) vector bundles. The action of $\mathbb{Z}/p$ in each fibre of $\nu_j(F)$ is multiplication by $\exp(2\pi ij/p)$. Classifying each of these bundles for each component of the fixed point set gives a $\mathfrak{U}_*$ module homomorphism

$$\nu : \mathfrak{U}_*^{\mathbb{Z}/p} \to \bigoplus_{\{d_j \ge 0\}} \mathfrak{U}_*(\Pi_{j=1}^{p-1} BU(d_j)) = \mathcal{M}.$$

This ring $\mathcal{M}$ is graded by saying that if $x \in \mathfrak{U}_n(\Pi\ BU(d_j))$ then $\deg x = n + \Sigma 2d_j$. There is also the notion of dimension of an element $x \in \mathcal{M}$. This is defined by saying that if $x \in \mathfrak{U}_n(\Pi\ BU(d_j))$ then $x = n + \Sigma 2(d_j - 1)$.

The ring $\mathcal{M}$ is isomorphic to the polynomial ring

$$\mathfrak{U}_*[\{\eta_{k,j};\ k \ge 1,\ 1 \le j \le p-1\}],$$

where $\eta_{k,j}$ denotes the equivariant normal bundle of

$CP^{k-1} = \{[z_0; z_1; \ldots; z_{k-1}; 0]\}$ in $CP^k(0, 0, \ldots, 0, j)$. In this notation the grading is deg $\eta_{k,j} = 2k$ and deg M = dim M if $M \in \mathfrak{U}_*$ while dim $\eta_{k,j} = 2(k-1)$ and dim M = dim M (the usual dimension) if $M \in \mathfrak{U}_*$.

Thus, if M is a $\mathbb{Z}/p$ connected unitary $\mathbb{Z}/p$ manifold of dimension m then $\nu(M)$ is a homogeneous polynomial of degree m, while dim $\nu(M)$ is the maximum dimension, up to bordism, of the fixed point set of M.

DEFINITION 3.2. If $i \neq 0 \bmod p$ then define

x_i to be $\eta_{n,j}$ where $n = [i/p] + 1$ and $j = i - p[i/p]$.

If $i = kp$ then define x_i to be M_k.

Thus $\mathcal{M}$ is isomorphic to the polynomial ring $\mathbb{Z}[x_1, x_2, x_3, \ldots]$. Note that deg $x_i = [(i+p-1)/p]$ and dim $x_i = [i/p]$.

LEMMA 3.3. $\nu(M_n) = x_{np}$.

$$\nu(CP^{n+1}(0, 0, \ldots, 0, j)) = x_{np+j} + (x_j)^{n+1}.$$

$$\nu(R_j) = x_j + r_j x_{p-1} \quad \text{for} \quad (p+1)/2 \leq j \leq p-1.$$

PROOF. Direct calculation.

The map $\nu : \mathfrak{U}_*^{\mathbb{Z}/p}(r) \to \mathcal{M}$ factors through $\mathbb{Z}[\{x_i; 1 \leq i \leq rp-1\}] \subset \mathcal{M} = \mathbb{Z}[\{x_i; i \geq 1\}]$. By ordering the variables $\{x_i\}$ so that $x_i > x_j$ if $i > j$ we see immediately from lemma 3.3 that the image of ν (restricted to $\mathfrak{U}_*^{\mathbb{Z}/p}(r)$) contains

$$x_i + \text{lots}, \quad \text{for} \quad 1 \leq i \leq p-2 \quad \text{or} \quad p \leq i \leq rp-1$$

and

$$px_{p-1}.$$

Thus the image of ν contains terms of the following form

$$x_I + \text{lots}$$

$$(px_{p-1})^{\ell} x_I + \text{lots}$$

where $I = (i(1),\ i(2), \ldots, i(k))$ with $i(j) \neq p-1$ for $j = 1, 2, \ldots, k$ and $x_I = x_{i(1)}\ x_{i(2)} \ldots x_{i(k)}$. We shall shortly show that the image contains terms of the form

$$p^m (x_{p-1})^{\ell} x_I + \text{lots}$$

where $I = (i(1),\ i(2), \ldots, i(k))$ with $i(j) \neq p-1$ for $j = 1, 2, \ldots, k$ and $m = [(\ell + \deg x_I - |I| + p-2)/(p-1)]$. $(|I| = i(1) + i(2) + \ldots + i($

THEOREM 3.4. *Suppose that* $L = (\ell(1),\ \ell(2), \ldots, \ell(k))$ *and* $J = (j(1),\ j(2), \ldots, j(k))$ *are k-tuples of non-negative integers. If* $\mathcal{P}$ *is the set of subsets of the set* $\{(i,j) \in \mathbb{Z} \times \mathbb{Z};\ 1 \leq i \leq k,\ 0 \leq j \leq$ *then*

$$\nu(\Lambda_L\ CP^k(J)) = \sum_{\mathbf{S} \in \mathcal{P}} (-1)^{|\mathbf{S}|} (x_{p-1})^{|L|}\ \nu(CP^k(J - \mathbf{S}))$$

where $J - \mathbf{S}$ *is the k-tuple* $(j(1) - s(1),\ j(2) - s(2), \ldots,\ j(k) - s(k))$ *with* $s(i)$ *denoting the cardinality of the set* $\{(i,j);\ 0 \leq j \leq \ell(i),\ (i,j) \in \mathbf{S}.\}$ *Finally* $|\mathbf{S}| = s(1) + s(2) + \ldots + s(k)$ *is the cardinality of* $\mathbf{S}$.

PROOF. $\Lambda_L\ CP^k(J)$ is the set of equivalence classes

$$\{[z_o;z_1;\ldots;z_k] \times \prod_{i=1}^{k} \prod_{j=1}^{\ell(i)} (u_{i,j}, v_{i,j});[z_o;z_1;\ldots;z_k] \in CP^k, (u_{i,j}, v_{i,j}$$

where $[z_o;\ldots;z_k] \times \Pi\Pi(u_{i,j},\ v_{i,j}) = [z_o';\ldots;z_k'] \times \Pi\Pi(u_{i,j}',\ v_{i,j}$ if and only if there exists $c \in C^* \setminus \{0\}$ and $t_{i,j} \in S^1$ $(1 \leq i \leq k,$ $0 \leq j \leq \ell(i))$ such that $z_i' = cz_i \prod_j t_{i,j}$ for $1 \leq i \leq k$,

$z_o' = cz_o$, $u_{i,j}' = t_{i,j}\, u_{i,j}$ and $v_{i,j}' = t_{i,j}^{-1}\, v_{i,j}$.

The action of $\xi \in \mathbb{Z}/p \subset S^1$ is given by

$$[z_o;\, z_1;\, \dots;\, z_k] \times \Pi\Pi(u_{i,j},\, v_{i,j})$$

$$\to [z_o;\, z_1\, \xi^{j(1)};\, \dots;\, z_k\, \xi^{j(k)}] \times \Pi\Pi(u_{i,j},\, \xi^{-1}\, v_{i,j}).$$

Given any element $x \in M$, there is a set $\mathbf{S} \in \mathcal{P}$ such that $v_{i,j} \neq 0$ for $(i,j) \in \mathbf{S}$ and $u_{i,j} \neq 0$ for $(i,j) \notin \mathbf{S}$. If x is in the fixed point set then

$$t_{i,j} = \xi \quad \text{for} \quad (i,j) \in \mathbf{S}, \text{ and}$$

$$t_{i,j} = 1 \quad \text{for} \quad (i,j) \notin \mathbf{S}.$$

Furthermore

$$u_{i,j} = 0, \quad v_{i,j} = 1 \quad \text{for} \quad (i,j) \in \mathbf{S}, \text{ and}$$

$$u_{i,j} = 1, \quad v_{i,j} = 0 \quad \text{for} \quad (i,j) \notin \mathbf{S}.$$

The action of $\xi \in \mathbb{Z}/p$ on points x with $\{(u_{i,j},\, v_{i,j})\}$ as just given looks like

$$[z_o;\, z_1;\, \dots;\, z_k] \times \prod\prod_{(i,j)\in\mathbf{S}} (0,\ 1) \times \prod\prod_{(i,j)\notin\mathbf{S}} (1,\ 0)$$

$$\to [z_o;\, z_1\, \xi^{j(1)-s(1)};\, \dots;\, z_k\, \xi^{j(k)-s(k)}]$$

$$\times \prod\prod_{(1,j)\in\mathbf{S}} (0,\ 1) \times \prod\prod_{(i,j)\notin\mathbf{S}} (1,\ 0).$$

Thus we can easily see that the normal bundle of the components containing such fixed points is given by

$$(-1)^{|\mathbf{S}|}\, (x_{p-1})^{|L|}\, \nu(CP^k(J - \mathbf{S}))$$

with the factor $(-1)^{|\mathbf{S}|}$ appearing for orientation reasons.

It is not too difficult to formulate and prove the corresponding result for $\nu(\lambda_\ell(M, \Phi))$ - this is left to the reader.

The next two results follow from Theorem 3.4.

COROLLARY 3.5. *If* $L = (\ell(1), \ell(2), \ldots, \ell(k))$ *satisfies* $0 \le \ell(i) \le p-1$ *for* $i = 1, 2, \ldots, k$ *then*

$$\nu(\Lambda_L\ CP^k(0, 0, \ldots, 0)) = (x_{p-1})^{|L|} CP^k + \text{lots}.$$

COROLLARY 3.6. *Suppose* j *is an integer such that* $1 \le j \le p-1$. *If* $L = (\ell(1), \ell(2), \ldots, \ell(k+1))$ *satisfies* $0 \le \ell(i) \le p-1$ *for* $i = 1, 2, \ldots, k$ *and* $0 \le \ell(k+1) < j$ *then*

$$\nu(\Lambda_L\ CP^{k+1}(0, 0, \ldots, 0, j)) = x_{p-1}{}^{|L|}\ x_{kp+j} + \text{lots}.$$

We can now complete the proof of Theorem 3.1. If $i \not\equiv 0 \bmod p$ then the result follows from Corollary 3.6. If $i = kp$ then we know (from Corollary 3.5) that we have elements $(x_{p-1})^\ell CP^k$ + lots for $0 \le \ell \le i - \deg x_i$. Now, if $k < p-1$ then $CP^k = a_k\ M_k + \text{lots} = a_k\ x_{kp}$ + lots where $a_k \not\equiv 0 \bmod p$. If $\ell \ge 1$ then there is an integer $b_{k,\ell}$ such that $a_k b_{k,\ell} = 1 + c_{k,\ell}\ p^\ell$ for some $c_{k,\ell}$. Thus

$$\nu(b_{k,\ell}\ \Lambda_L\ CP^k(0, 0, \ldots, 0) - c_{k,\ell}(R_{p-1})^\ell\ M_k)$$

$$= (x_{p-1})^\ell\ x_{kp} + \text{lots}$$

where $\ell = |L|$. (Recall that $\nu(R_{p-1}) = px_{p-1}$.)

4. ISOLATED FIXED POINTS

In this section we shall concentrate on the manifolds appearing in the statement of Theorem 2.1. The aim will be to prove the following result.

THEOREM 4.1. _There exist_ $\mathbb{Z}/p$ _manifolds_ $N(n, j)$ _for_ $n = 1, 2, \ldots, p-1$ _and_ $n \leq j \leq p-1$, _such that_

(i) $N(n, j)$ _is obtainable from the "generators" of_ $\mathfrak{U}_*^{\mathbb{Z}/p}(1)$,

(ii) $\dim N(n, j) = 2n$, _and_

(iii) $$\nu(N(n, j)) = \begin{cases} (y_{p-1})^{n-1}\, y_j + \text{lots if } j \leq p-2 \\ \\ p(y_{p-1})^n + \text{lots if } j = p-1. \end{cases}$$

The y_i _are variables such that_ $\mathbb{Z}[x_1, x_2, \ldots, x_{p-1}] \cong \mathbb{Z}[y_1, y_2, \ldots, y_{p-1}]$ _with a (non-obvious) ordering_ $y_{p-1} > y_{p-3} > y_{p-5} > \ldots > y_2 > y_{p-2} > y_{p-4} > \ldots > y_1$.

The proof of this result will occupy the whole of this section. The next lemma is in part Lemma 3.3

LEMMA 4.2.

(i) $\nu(R_j) = x_j + r_j x_{p-1}$ _for_ $(p+1)/2 \leq j \leq p-1$, _where_ $jr_j = 1 \bmod p$ _and_ $1 < r_j < p$.

(ii) $\nu(CP^1(p-j)) = x_j + x_{p-j}$ _for_ $1 \leq j \leq p-1$,

(iii) $\nu(CP^2(j,1)) = x_1\, x_j + x_{p-1}\, x_{j-1} + x_{p-j}\, x_{p-j+1}$ _for_ $2 \leq j \leq p-1$.

PROOF. Direct calculation.

We now introduce the variables $\{P_1, P_2, \ldots, P_{(p-1)/2}, Q_{(p+1)/2}, \ldots, Q_{p-2}, y\}$ which up to order will be the set of variables $\{y_1, y_2, \ldots, y_{p-1}\}$.

DEFINITION 4.3.

(i) Define P_j to be $x_j + x_{p-j}$ for $1 \leq j \leq (p-1)/2$.

(ii) Let q_j be the unique integer satisfying

$((p^{p-1}-p)/2 + j)q_j \equiv 1 \bmod p^{p-1}$ and $1 < q_j < p^{p-1}$.

Define Q_j to be $x_j + q_j\, x_{p-1}$ for $(p+1)/2 \le j \le p-2$.

(iii) Define y to be x_{p-1}.

There are technical reasons for defining q_j in the above way. These reasons should become apparent towards the end of this section.

REMARK 4.4. $\mathbb{Z}[x_1, x_2, \ldots, x_{p-1}] \cong \mathbb{Z}[P_1, P_2, \ldots, P_{(p-1)/2}, Q_{(p+1)/2}, \ldots, Q_{p-2}, y]$.

REMARK 4.5. $r_j - q_j$ is divisible by p.

LEMMA 4.6.

(i) $\nu(CP^1(p-j) = \nu(CP^1(j)) = P_j$ <u>for</u> $1 \le j \le (p-1)/2$.

(ii) $\nu(R_j - ((q_j - r_j)/p)R_{p-1}) = Q_j$ <u>for</u> $(p+1)/2 \le j \le p-2$.

(iii) $\nu(R_{p-1}) = py$.

In other words the image of $\nu : \mathfrak{U}_*^{\mathbb{Z}/p}(1) \to \mathbb{Z}[P_1, \ldots, Q_{p-2}, y]$ contains the subring generated by $P_1, \ldots, Q_{p-2}$.

If we order $\{P_j\}$ as $P_1 > P_2 > \ldots > P_{(p-1)/2}$ then we have the following result.

LEMMA 4.7. <u>If</u> $1 \le j \le (p-1)/2$ <u>and</u> $0 \le \ell \le p-1 - 2j$ <u>then</u>

$$\nu(\lambda_\ell CP^1(p-j)) = y^\ell P_j + \text{lots}.$$

PROOF. $\lambda_\ell CP^1(p-j) = \Lambda_\ell CP^1(p-j)$ and so by Theorem 3.4

$$\nu(\lambda_\ell CP^1(p-j)) = \sum_{s=0}^{\ell} (-1)^s \binom{\ell}{s} y^\ell \nu(CP^1(p-j-s))$$

$$= \sum_{s=0}^{\ell} (-1)^s \binom{\ell}{s} y^\ell P_{j+s} + \sum_{s=m+1}^{\ell} (-1)^s \binom{\ell}{s} y^\ell P_{p-j-s}$$

where $m = \min\{(p-1)/2 - j, \ell\}$. The highest order term is $y^\ell P_j$ if $j < p - j - \ell$ i.e. if $\ell < p-2j$. If $j = p-j-\ell$ then ℓ is odd (p is odd) and $\nu(\lambda_\ell CP^1(p-j)) = 0$.

The proofs of the next few lemmata are by straightforward calculation and are left for the reader.

LEMMA 4.8. *If* $(p+1)/2 < j < p-1$ *then*

$$\nu(CP^2(j,\ 1)) = -(1 + q_{j-1})y\ Q_j + (1-q_j)y\ Q_{j-1}$$
$$+ q_j\ y\ P_{p-j+1} + q_{j-1}y\ P_{p-j} - q_j y\ P_1$$
$$+ P_1\ Q_j + (P_{p-j+1} - Q_{j-1})(P_{p-j} - Q_j)$$
$$+ (q_j - q_{j-1} + q_{j-1}\ q_j)y^2.$$

Note that $1 + q_{j-1} = j/(j-1) \bmod p$

$\neq 0 \bmod p.$

Note also that $q_j - q_{j-1} + q_{j-1}\ q_j = 0 \bmod p^{p-1}$.

LEMMA 4.9.

$$\nu(CP^2(p-1,1)) = -(2q_{p-2} + 1)y^2 + 2y\ Q_{p-2} - y\ P_2$$
$$+ (1 + q_{p-2})y\ P_1 + P_1(P_2 - Q_{p-2}).$$

Observe that $(4-p)(2q_{p-2} + 1) = -p \bmod p^{p-1}$.

LEMMA 4.10. *If* $h = (p+1)/2$ *then*

$$\nu(CP^2(h,1)) = -(2 - q_h + q_{h-1})y\ Q_h + (1 + q_{h-1})y\ P_{h-1}$$
$$- q_h\ y\ P_1 + P_1\ Q_h$$
$$+ (P_{h-1} - Q_h + (q_h + q_{h-1})y)(P_{h-1} - Q_h)$$
$$+ (2q_h - q_h\ q_h)y^2.$$

Note that $q_h = 2$, $q_{h-1} = p^{p-1}-2$ and so $-(2-q_h+q_{h-1}) = 2-p^{p-1}$

while $q_h + q_{h-1} = p^{p-1}$ and $2q_h - q_h q_h = 0$. In other words

$$\nu(CP^2(h,1)) = (2-p^{p-1})y\ Q_h + (p^{p-1}-1)y\ P_{h-1} - 2y\ P_1 + P_1\ Q_h$$

$$+ (P_{h-1} - Q_h + p^{p-1}y)(P_{h-1} - Q_h).$$

LEMMA 4.11. *If* $2 < j \leq (p-1)/2$ *then*

$$\nu(CP^2(j,1)) = -(1+q_{p-j})y\ Q_{p-j+1} + (1-q_{p-j+1})y\ Q_{p-j}$$

$$+ P_1(P_j - Q_{p-j}) + y\ P_{j-1} + q_{p-j}y\ P_1 - y\ P_j$$

$$+ Q_{p-j+1}\ Q_{p-j} + (q_{p-j+1} - q_{p-j} + q_{p-j+1}\ q_{p-j})y^2.$$

LEMMA 4.12. $\nu(CP^2(2,1)) = \nu(CP^2(p-1,1))$.

By ordering the variables $\{y,\ P_i,\ Q_i\}$ so that $y > Q_{p-2} > \ldots > Q_{(p+1)/2} > P_1 > P_2 > \ldots > P_{(p-1)/2}$ we can state the above results in the following form.

PROPOSITION 4.13. *Let* $q_j - q_{j-1} + q_{j-1}\ q_j = t_j\ p^{p-1}$.

(i) $\nu(CP^2(p-1,\ 1)) = -(2q_{p-2} + 1)y^2 + \text{lots}$.

(ii) *If* $(p+1)/2 < j < p-1$ *then*

$$\nu(CP^2(j,1)) = -(1+q_{j-1})y\ Q_j + t_j\ p^{p-1}\ y^2 + \text{lots}.$$

(iii) *If* $h = (p+1)/2$ *then*

$$\nu(CP^2(h,1)) = (2-p^{p-1})y\ Q_h + \text{lots}.$$

(iv) *If* $2 < j \leq (p-1)/2$ *then*

$$\nu(CP^2(j,1)) = -(1+q_{p-j})y\ Q_{p-j+1} + t_{p-j}\ p^{p-1}\ y^2 + \text{lots}.$$

(v) $\nu(CP^2(2,1)) = -(2q_{p-2} + 1)y^2 + \text{lots}$.

LEMMA 4.14. *There exist* $\mathbb{Z}/p$ *manifolds* $M(n,\ j)$ *for* $j = (p+1)/2, \ldots, p-1$ *and* $n = 2,3,\ldots,2j-p+1$ *such that*

(i) $M(n, j)$ <u>is obtainable from the "generators" of</u> $\mathfrak{U}_*^{\mathbb{Z}/p}(1)$.

(ii) $\dim M(n, j) = 2n$, <u>and</u>

(iii) $$\nu(M(n,j)) = \begin{cases} y^{n-1} Q_j + \text{lots} & \underline{\text{if}} \quad (p+1)/2 \leq j < p-1. \\ py^n + \text{lots} & \underline{\text{if}} \quad j = p-1. \end{cases}$$

PROOF. For brevity let us denote

$-(1+q_{j-1})y\,Q_j + t_j\,p^{p-1}y^2$ by D_j, $(h < j < p-1)$,

$-(2q_{p-2} + 1)y^2$ by D_{p-1}, and

$(2 - p^{p-1})y\,Q_h$ by D_h

where $h = (p+1)/2$ as before. Since $\lambda_\ell\, CP^2(j, 1) = \Lambda_{\ell,0}\, CP^2(j,1)$ we have

$$\nu(\lambda_\ell\, CP^2(j,1)) = \sum_{s=0}^{\ell} (-1)^s \binom{\ell}{s} y^\ell\, \nu(CP^2(j-s,1))$$

$$= \sum_{s=0}^{\ell} (-1)^s \binom{\ell}{s} y^\ell D_{j-s} +$$

$$\sum_{s=m+1}^{\ell} (-1)^s \binom{\ell}{s} y^\ell D_{p-j+s+1} + \text{lots}$$

where $m = \min\{j-h, \ell\}$. The highest term is D_j if $j > p-j+\ell+1$, i.e. if $\ell > 2j - p-1$. If $\ell = 2j - p-1$ then ℓ is even and so the highest term is $2D_j$.

Let $\ell_j = 2$ if $\ell = 2j - p-1$ and 1 otherwise, then the highest term is $\ell_j D_j$. Let $m_{\ell,j}$, $k_{\ell,j}$ be integers such that

$-m_{\ell,j}(1+q_{j-1})\ell_j = 1 + k_{\ell,j}\, p^{p-1}$ for $h < j < p-1$,

$-m_{\ell,p+1}(2q_{p-2} + 1)\ell_{p-2} = p + k_{\ell,p-1}\, p^{p-1}$, and

$m_{\ell,h}(2 - p^{p-1})\ell_h = 1 + k_{\ell,h}\, p^{p-1}$.

The manifolds M(n,j) may now be defined as follows:

$$M(n,p-1) = m_{n-2,p-1}\,\lambda_{n-2}\,CP^2(p-1,1) - k_{n-2,p-1}\,p^{p-n-1}(R_{p-1})^n.$$

$$M(n,j) = m_{n-2,j}\,\lambda_{n-2}\,CP^2(j,1) - k_{n-2,j}\,p^{p-n}(R_{p-1})^{n-1}\,\bar{Q}_j$$

$$-m_{n-2,j}\,\ell_j\,t_j\,p^{p-n-1}(R_{p-1})^n, \quad \text{if} \quad h < j < p-1.$$

$$M(n,h) = m_{n-2,h}\lambda_\ell CP^2(h,1) - k_{n-2,h}\,p^{p-n}(R_{p-1})^{n-1}\,\bar{Q}_h.$$

The manifolds $\bar{Q}_j$ satisfy $\nu(\bar{Q}_j) = Q_j$ - see lemma 4.6. This completes the proof of lemma 4.14.

To prove the main assertion of this section - i.e. Theorem 4.1 - we let

$$N(n,\ p+1-2j) = M(n,\ p-j) \quad \text{for} \quad 1 \le j \le h-1,$$

$$N(n,\ p-2j) = \lambda_{n-1}\,CP^1(p-j) \quad \text{for} \quad 1 \le j \le h-1,$$

$$y_{p-1} = x_{p-1} = y,$$

$$y_{p-1-2j} = Q_{p-j} \quad \text{for} \quad 1 < j \le h-1,$$

$$y_{p-2j} = P_j \quad \text{for} \quad 1 \le j \le h-1.$$

The ordering $y > Q_{p-2} >\ldots> Q_h > P_{h-1} >\ldots> P_1$ induces the ordering $y_{p-1} > y_{p-3} >\ldots> y_2 > y_{p-2} > y_{p-4} >\ldots> y_1$ and this completes the proof of theorem 4.1.

5. INTEGRALITY CONDITIONS

The purpose of this section is to describe necessary conditions for a homogeneous polynomial $P \in \mathfrak{M}$ to lie in the image

$\nu : \mathfrak{U}_*^{\mathbb{Z}/p} \to \mathcal{M}$. An obvious inner product $\langle\ ,\ \rangle$ is defined on $\mathcal{M}$, (see paragraph following Definition 5.4). Elements $S_J \in \mathcal{M}$, (J an unordered k-tuple, $k \geq 0$) are found such that if $M \in \mathfrak{U}_{2n}^{\mathbb{Z}/p}(r)$ and $\nu(M) = p^{\ell}R$ then $\langle S_J, R\rangle = 0 \bmod p^{\ell+1}$ for all J satisfying $|J| < n-\ell(p-1)$. See Theorem 5.5. Different k-tuples sometimes produce linearly dependent elements, Theorem 5.7 describes a set of linearly independent elements.

Let M be a 2n dimensional $\mathbb{Z}/p$ manifold and let F be a 2d dimensional component of the fixed point set. Split the normal bundle of F in M as in section 3, i.e. $\nu(F, M) = \Sigma\, \nu_j(F)$. For each j, $1 \leq j \leq p-1$, denote by $z_{j,1}(F), j_{,2}(F), \ldots, z_{j,d(j)}(F)$ the formal symbols whose elementary symmetric functions give the Chern classes of $\nu_j(F)$, ($d(j)$ = complex dimension of $\nu_j(F)$). For brevity denote the set $z_{1,1}(F), \ldots, z_{1,d(1)}(F), \ldots, z_{p-1,1}(F), \ldots, z_{p-1,d(p-1)}(F)$ by $z_1, z_2, \ldots, z_{n-d}$. Denote the corresponding set.

$$\underbrace{1,1,\ldots,1}_{d(1)},\underbrace{2,2,\ldots,2}_{d(2)},\ldots,\underbrace{p-1,p-1,\ldots,p-1}_{d(p-1)}$$

by $t_1, t_2, \ldots, t_{n-d}$. (Note that if $z_k = z_{i,j}(F)$ then $t_k = i$.) Finally, let $y_1, y_2, \ldots, y_{n-d}$ be the formal symbols whose elementary symmetric functions give the Chern classes of F.

DEFINITION 5.1. Let f be a symmetric homogeneous polynomial in n variables. Define $f(e_p(F))$ to be the mod p reduction of the integer

$$\{f(\ldots, y_j, \ldots, \ldots, z_j + t_j, \ldots)\, \Pi(z_j + t_j)^{-1}\}[F].$$

THEOREM 5.2. *If* $M \in \mathfrak{U}_{2n}^{\mathbb{Z}/p}(r)$, *if* $1 \leq r \leq p-1$ *and if* $\nu(M)$ *is divisible by* p^{ℓ} *then*

$$\sum_{F \in \mathrm{Fix}} f(e_p(F)) = 0 \text{ mod } p^{\ell+1}$$

for all symmetric homogeneous polynomials f of degree less than $n - \ell(p-1)$.

PROOF. Define $f(e(F))$ to be the element of $Q[\xi]$, where $\xi = \exp(2\pi i/p)$, given by

$$\{f(\ldots, \exp(y_j)-1, \ldots, \ldots, \exp(z_j + 2\pi i t_j/p)-1, \ldots)$$

$$\Pi(\exp(z_j + 2\pi i t_j/p)-1)^{-1} \, \Pi(y_j/(\exp(y_j)-1))\}[F].$$

We then have the well known integrality condition

$$\Sigma f(e(F)) \in \mathbb{Z}[\xi]$$

where the summation is over the components F of the fixed point set. This integrality condition may be obtained quite easily, for example, from the K theory localisation theorem [1].

Multiplying the integrality condition by $(\xi-1)^{n-\ell(p-1)-d(f)}$, where $d(f)$ is the degree of f, and reducing modulo $(\xi-1)$ produces the desired result. The case $r = 1$, general ℓ is proved in Proposition 4.1 of [4]. The case $\ell = 0$, $0 \leq r \leq p-1$ is given as Corollary 2.5 of [2]. See also Theorems 1.1 and 1.2 in [6].

The above result will be restated in a different form shortly. The ring $\mathbb{Z}[x_1, x_2, \ldots, x_{rp-1}]$ is graded, $\mathbb{Z}[\{x_i\}] = \oplus \mathcal{M}(n, r)$ where $\mathcal{M}(n, r)$ consists of those homogeneous polynomials in $\mathbb{Z}[\{x_i\}]$ of degree n. (Alternatively, $\mathcal{M}(n, r)$ consists of those elements in $\mathcal{M}$ of degree n and dimension less than or equal to r.)

DEFINITION 5.3. Let $\mathfrak{R}(n, r)$ be the set of unordered sequences integers $I = (i(1), i(2), \dots, i(k))$, $k \geq 0$ such that $1 \leq i(j) \leq rp-1$ for $j = 1, 2, \dots, k$ and $\deg I = \Sigma_{j=1}^{k}[(i(j) + p-1)/p] =$

Thus $\mathfrak{M}(n, r)$ is a $\mathbb{Z}$ module with basis $\{x_I ; I \in \mathfrak{R}(n, r)\}$ where x_I denotes, in the usual fashion, $x_{i(1)} x_{i(2)} \cdots x_{i(k)}$ if $I = (i(1), i(2), \dots, i(k))$.

Suppose that $J = (j(1), j(2), \dots, j(k))$ is an unordered sequence of positive integers of length $k \leq n$. Define $s_J(z_1, z_2, \dots, z_n$ to be the smallest symmetric homogeneous polynomial in the variables $z_1, z_2, \dots, z_n$ that contains the monomial $z_1^{j(1)} z_2^{j(2)} \cdots z_k^{j(k)}$.

DEFINITION 5.4. Define $S_J \in \mathfrak{M}(n, r)$ to be any lift of the following vector in $\mathfrak{M}(n, r) \otimes \mathbb{Z}/p$:

$$\sum_{I \in \mathfrak{R}(n,r)} s_J[x_I] x_I$$

where $s_J[x_I]$ means $\Sigma\, s_J(e_p(F))$, $F \in \mathrm{Fix}(x_I)$.

By defining the obvious inner product $\langle\, , \rangle$ on $\mathfrak{M}(n, r)$, i.e. $\langle \Sigma\, a_I\, x_I, \Sigma\, b_I\, x_I \rangle = \Sigma\, a_I\, b_I$, we can restate Theorem 5.2 in the following way.

THEOREM 5.5. (<u>Restatement of Theorem 5.2</u>). <u>If</u> $M \in \mathfrak{U}_{2n}^{\mathbb{Z}/p}(r)$, <u>if</u> $1 \leq r \leq p-1$ <u>and if</u> $\nu(M)$ <u>is divisible by</u> p^{ℓ} <u>then</u> $\langle S_J, \nu(M) \rangle = 0$ <u>mod</u> $p^{\ell+1}$ <u>for all J such that</u> $|J| < n-\ell(p-1)$.

As was mentioned previously, different k-tuples J, J′ may produce linearly dependent elements S_J, $S_{J'}$. A set of linearly independent elements is described after the next definition.

DEFINITION 5.6. Let $\mathbf{S}(n, r)$ be the set of unordered k-tuples $J = (j(1), j(2), \dots, j(k))$ for $0 \leq k \leq n$ satisfying $1 \leq j(i) \leq rp-1$ and $j(i) \neq p-1$ for $i = 1, 2, \dots, k$.

THEOREM 5.7. <u>The set of elements</u> $\{S_J \in \mathcal{M}(n,r),\ J \in \mathbf{S}(n,r)\}$ <u>are linearly independent over</u> $\mathbb{Z}/p$.

The proof of this theorem will occupy the rest of this section.

We shall need to know $s_J[x_I]$ for all $I \in \mathcal{R}(n,r)$ and $J \in \mathbf{S}(n,r)$. Because of the multiplicative properties of s_J, i.e. $s_J[x_A\ x_B] = \sum_{K \cup L = J} s_K[x_A]\ s_L[x_B]$, we need only calculate $s_J[x_i]$ for all $J \in \mathbf{S}(n,r)$ and $i = 1, 2, \ldots, rp-1$, Unfortunately $s_J[x_i] \neq 0$ if $J = (j(1), j(2), \ldots, j(k))$ with $k > 1$ which makes the calculation of $s_J[x_I]$ extremely difficult. We circumvent this problem by replacing the set $\{x_i\}$ by $\{X_i\}$, (definition 5.10), so that $s_J[X_i] = 0 \bmod p$ if $J = (j(1), j(2), \ldots, j(k))$ with $k > 1$. The set $\{X_i\}$ is defined in terms of the elements N_n and $\mu_{n,k}$ described in the next definition.

DEFINITION 5.8. Define N_n, $\mu_{n,k} \in \mathcal{M} \otimes \mathbb{Z}/p$ for $0 \le n \le p-2$, $1 \le k \le p-1$ inductively by

$$N_o = \text{point}$$

$$N_n = (n+1)^{-1}(CP^n - \sum_{I, I \neq (n)} s_I[CP^n] N_I).$$

$$\mu_{o,k} = k\ \eta_{1,k}.$$

$$\mu_{n,k} = k^{n+1}\ \eta_{n+1,k} - \sum_{I, I \neq (0)} s_I[CP^n] k^{|I|} N_I \mu_{n-|I|,k}.$$

The symbol I denotes unordered sequences of positive integers $(i(1), i(2), \ldots, i(\ell))$ and N_I denotes (as usual) the product $N_{i(1)} N_{i(2)} \cdots N_{i(\ell)}$. The terms $s_I[CP^n]$ and $\overline{s}_I[CP^n]$ are given by

$$\{s_I(y_1, y_2, \ldots, y_n)\}[CP^n] \text{ and}$$

$$\{s_I(y_1, y_2, \ldots, y_n)z^{n-|I|}\}[CP^n]$$

respectively, where elementary symmetric functions of $y_1, y_2, \ldots, y_n$ give the Chern classes of CP^n and z is the first Chern class of the normal bundle of CP^n in CP^{n+1} (i.e. of the conjugate Hopf bundle).

THEOREM 5.9.

$$s_J[N_n] = \begin{cases} 1 \bmod p & \underline{\text{if}} \quad J = (n) \\ 0 \bmod p & \underline{\text{otherwise}} \end{cases}$$

$$s_J[N_I] = \begin{cases} 1 \bmod p & \underline{\text{if}} \quad J = I \quad (\underline{\text{up to order}}) \\ 0 \bmod p & \underline{\text{otherwise}} \end{cases}$$

$$s_J[\mu_{n,k}] = \begin{cases} (-1)^n \bmod p & \underline{\text{if}} \quad j = (0) \\ \binom{t-1}{n}k^t \bmod p & \underline{\text{if}} \quad J = (t) \ \underline{\text{and}} \ t > n \\ 0 \bmod p & \underline{\text{otherwise}} \end{cases}$$

The proof of Theorem 5.9 is by straightforward (but long) induction and is left to the reader.

DEFINITION 5.10. Suppose $i = np + k$ where $0 \le k \le p-1$, define X_i by

$$X_i = \begin{cases} N_n & \text{if} \quad k = 0 \\ \mu_{n,k} & \text{if} \quad k \neq 0. \end{cases}$$

COROLLARY 5.11.

$s_J[X_i] = 0 \bmod p$ <u>if</u> $J = (j(1), j(2), \ldots, j(\ell))$ <u>with</u> $\ell > 0$.

$$s_j[X_i] = \begin{cases} 1 \bmod p & \underline{\text{if}} \quad i = jp \\ (-1)^n \bmod p & \underline{\text{if}} \ j = 0 \ \underline{\text{and}} \ i = np+k \\ \binom{j-1}{n}k^j \bmod p & \underline{\text{if}} \ i = np+k \ \underline{\text{and}} \ j > n \\ 0 \bmod p & \underline{\text{otherwise}}. \end{cases}$$

Observe that

$$s_o[X_{np+k}] = \begin{cases} 0 \bmod p & \text{if } k = 0 \\ (-1)^n \bmod p & \text{if } k \neq 0 \end{cases}$$

and

$$s_{p-1}[X_{np+k}] = \begin{cases} 0 \bmod p & \text{if } k = 0 \\ \binom{p-2}{n} k^{p-1} \bmod p & \text{if } k \neq 0. \end{cases}$$

But $\binom{p-2}{n} k^{p-1} = \binom{p-2}{n} = (-1)^n (n+1) = (-1)^n [(np + k + p-1)/p] \bmod p$. In other words $s_{p-1}[X_j] = \deg(X_j)\, s_o[X_j]$, which explains why the condition $j(i) \neq p-1$ was included in the definition of $\mathbf{S}(n,r)$ in 5.6, (i.e. if $J = (j(1), j(2),\ldots,j(\ell))$ and $J' = (j(1) + (p-1)\,a(1),\ldots,$ $j(\ell) + (p-1)\,a(\ell),\ (p-1)\,a(\ell+1),\ldots,\ (p-1)\,a(m))$ then S_J and $S_J{}'$ are linearly dependent).

THEOREM 5.12. <u>The set of vectors of the form</u> $\Sigma_{i=1}^{rp-1}\ s_j[X_i]X_i \in (\mathbb{Z}/p)[X_1, X_2,\ldots,X_{rp-1}]$ <u>for</u> $0 \leq j \leq rp-1$ <u>and</u> $j \neq p-1$ <u>are linearly independent</u>.

PROOF. Define $f_j[X_i] \in \mathbb{Z}/p$ by

$$f_j[X_i] = \Sigma_{\ell=0}^{b} (-1)^{b-\ell} \binom{b}{\ell}\ s_{(b-\ell)(p-1)+b+c}[X_i]$$

where $j = bp+c$ with $0 \leq c < p$.

We shall now prove the following "claim"

$$f_{bp+c}[X_{np+k}] = \begin{cases} 1 & \text{if } b = n \text{ and } c = k = 0 \\ k^{b+c} & \text{if } b = n \text{ and } c > 0 \\ 0 & \text{if } b > n \\ ? & \text{otherwise} \end{cases}$$

If $k = 0$ then it is clear that $f_{bp+c}[X_{np}] = 1$ in case $b+c = n$ and 0 otherwise. For $k \neq 0$ we have

$$f_{bp+c}[X_{np+k}] = \sum_{\ell=0}^{b} (-1)^{b-\ell}\binom{b}{\ell} s_{(b-\ell)(p-1)+b+c}[X_{np+k}]$$

$$= \sum_{\ell=0}^{b} (-1)^{b-\ell}\binom{b}{\ell}\binom{(b-\ell)(p-1)+b+c-1}{n}_k (b-\ell)(p-1)+b+c$$

$$\text{if } b+c > n$$

$$= \sum_{\ell=0}^{b} (-1)^{b-\ell}\binom{b}{\ell}\binom{bp+c-1-(p-1)\ell}{n}_k k^{b+c} \pmod p$$

$$= k^{b+c} \sum(-1)^{b-\ell}\binom{b}{\ell}(\ell^n/n! + P_{n-1}(\ell))$$

where $P_{n-1}(\ell)$ is a polynomial in ℓ of degree less than or equal to $n-1$. The claim will follow from the following fact

$$\sum_{\ell=0}^{b} (-1)^{b-\ell}\binom{b}{\ell}\ell^n = \begin{cases} b! & \text{if } n = b \\ 0 & \text{if } 0 \le n < b. \end{cases}$$

To prove this fact note that the result is true if $n = 0$. For $n > 0$ the result follows by induction by noting that

$$\frac{d^n}{dx^n}(x-1)^b = b(b-1) \ldots (b-n+1)(x-1)^{b-n}$$

and

$$\frac{d^n}{dx^n}\left(\sum(-1)^{n-\ell}\binom{b}{\ell}x^\ell\right) = \sum(-1)^{n-\ell}\binom{b}{\ell}\ell(\ell-1) \ldots (\ell-n+1)x^{\ell-n}$$

$$= \sum(-1)^{n-\ell}\binom{b}{\ell}\ell^n + \sum(-1)^{n-\ell}\binom{b}{\ell} Q_{n-1}(\ell)$$

where $Q_{n-1}(\ell)$ is a polynomial of degree $\le n-1$. Putting $x = 1$ produces the desired result.

Consider $f_{bp+c}[X_{np+k}]$ as the coefficients of the matrix $M = (M_{bp+c,\,np+k})$. This matrix M has a "block upper triangular form" - the blocks are $\{m_{bp,\,np}\}$ where $(m_{bp,\,np})_{c,k} = M_{bp+c,\,np+k}$. The blocks $m_{bp,\,np}$ with $b > n$ are zero while the block $M_{bp,\,bp}$ has the form

$$(m_{bp,bp})_{c,k} = \begin{cases} 1 & \text{if } c = k = 0 \\ ? & \text{if } c = 0,\ k > 0 \\ 0 & \text{if } c > 0,\ k = 0 \\ k^{b+c} & \text{if } c > 0,\ k > 0 \end{cases}$$

which is clearly non-singular. Therefore M is non-singular and hence the vectors

$$\sum_i f_j[X_i]X_i$$

are linearly independent and hence so are the vectors

$$\sum_i s_j[X_i]X_i$$

since $\sum_i f_j[X_i]X_i = \sum_i s_j[X_i]X_i + \sum_{\ell<j} (\sum_i a_\ell\ s_\ell[X_i]X_i)$ for some a_ℓ.

COROLLARY 5.13. <u>The vectors</u> $\sum_{I \in \mathbf{R}(n,r)} s_J[X_I]X_I \in (\mathbb{Z}/p)[X_1, X_2, \ldots, X_{rp-1}]$ <u>for</u> $J \in \mathbf{S}(n,r)$ <u>are linearly independent</u>.

This follows from Theorem 5.12, the multiplicative properti[es] of s_J and Corollary 5.11.

Theorem 5.6. follows immediately because

$$X_i = a_i\ x_i + \text{lots mod } p$$

$$x_i = b_i\ X_i + \text{lots mod } p$$

with a_i, $b_i \neq 0$ mod p and $s_J[A+B] = s_J[A] + s_J[B]$.

6. PROOF OF THE MAIN RESULT

In this section we use the results of the previous sections to prove two theorems which when combined prove the main result.

In section 4 the terms $y_1, y_2, \ldots, y_{p-1}$ were defined together with some order $(y_{p-1} > y_{p-3} > \ldots > y_1)$. Define

$y_p, y_{p+1}, \ldots, y_{rp-1}$ to be $x_p, x_{p+1}, \ldots, x_{rp-1}$ respectively together with the ordering $y_{rp-1} > y_{rp-2} > \ldots > y_p > y_{p-1}$. From Theorem 3.1 and Theorem 4.1 we see that the image of ν contains the following elements.

$$(y_{p-1})^d y_j + \text{lots} \quad \text{for } j = 1,2,\ldots,p-2,p,p+1,\ldots,rp-1$$
$$\text{and } d = 0,1,\ldots,j\text{-deg } j.$$

$$p(y_{p-1})^d + \text{lots} \quad \text{for } d = 0,1,\ldots,p-1$$

where $\deg j = [(j+p-1)/p] = \deg y_j$.

Let $\mathcal{J}(n,r)$ denote the image of $\mathfrak{U}_*^{\mathbb{Z}/p}(r)$ in $\mathcal{M}(n,r) \subset \mathbb{Z}[y_1, y_2, \ldots, y_{rp-1}]$.

THEOREM 6.1. $\mathcal{J}(n,r)$ <u>contains the</u> $\mathbb{Z}$ <u>submodule of</u> $\mathcal{M}(n,r)$ <u>with basis</u>

$$p^{q(I)}(y_{p-1})^{n-\deg I}\, y_I + \text{lots}$$

<u>for</u> $I \in \mathbf{S}(n,r)$ <u>with deg</u> $I \le n$, <u>where</u> $q(I) = [(n - |I| + p-2)/(p-1)]$.

PROOF. Suppose that we are given $I = (i(1), i(2), \ldots, i(\ell)) \in \mathbf{S}(n,r)$ with $\deg I \le n$. If $n \le |I|$ then $q(I) = 0$, in this case we shall presently show that there exists $D = (d(1), d(2), \ldots, d(\ell))$ such that $0 \le d(j) \le i(j) - \deg i(j)$ and $|D| = n - \deg I$. If this is so then

$$\prod_{j=1}^{\ell} ((y_{p-1})^{d(j)}\, y_{i(j)} + \text{lots})$$

$$= (y_{p-1})^{n-\deg I}\, y_I + \text{lots}$$

belongs to $\mathcal{J}(n,r)$. To prove the existence of such D let $c(j) = i(j) - \deg i(j)$. So $c(j) \ge 0$ and

$$\sum_{j=0}^{\ell} c(j) = |I| - \deg I \ge n - \deg I \ge 0$$

therefore there is an integer k $(0 \le k \le p-1)$ such that

$$\sum_{j=1}^{k} c(j) \le n - \deg I, \text{ and}$$

$$\sum_{j=1}^{k+1} c(j) \ge n - \deg I.$$

Now define $D = (d(1), d(2), \ldots, d(\ell))$ by $d(j) = c(j)$ for $j = 1,2,\ldots,k$ $d(k+1) = n - \Sigma_{j=1}^{k} c(j) - \deg I$ and $d(j) = 0$ for $j = k+2, k+3, \ldots, \ell$. It is easy to see that D has the required properties.

If $n > |I|$ then define $D = (d(1), d(2), \ldots, d(\ell))$ by $d(j) = i(j) - \deg i(j)$ for $j = 1,2,\ldots,\ell$. Consider the following element

$$(p(y_{p-1})^{p-1} + \text{lots})^{a} \, (p(y_{p-1})^{b} + \text{lots}) \sum_{j=1}^{\ell} ((y_{p-1})^{d(j)} \, y_{i(j)} + \text{lots})$$

where $a = [(n - |I|)/(p-1)]$ and $b = n - |I| - (p-1)[(n - |I|)/(p-1)]$. Note in particular, that $b \le p-1$. This element lies in $\mathcal{J}(n,r)$ and has the form

$$p^{q(I)} \, (y_{p-1})^{n-\deg I} \, y_I + \text{lots}$$

since $a+b = q(I)$ and $a(p-1) + b + |D| = n - \deg I$. This completes the proof of Theorem 6.1.

THEOREM 6.2. <u>There exist</u> $\theta_1, \theta_2, \ldots, \theta_{rp-1}$ <u>in</u> $\mathbb{Z}[y_1, y_2, \ldots, y_{rp-1}]$ <u>together with some ordering such that</u> $\mathcal{J}(n,r)$ <u>is contained in the</u> $\mathbb{Z}$ <u>submodule of</u> $\mathcal{M}(n,r)$ <u>with basis</u>

$$p^{q(I)} (\theta_{p-1})^{n-\deg I} \, \theta_I + \text{lots}$$

<u>for</u> $I \in \mathbf{S}(n,r)$ <u>with</u> $\deg I \le n$, <u>where</u> $q(I) = [(n - |I| + p-2)/(p-1)]$.

PROOF. From Theorem 5.5 (restatement of Theorem 5.2) we know that if $P \in \mathcal{J}(n,r)$ and if P is divisible by p^k then

$\langle S_J, P\rangle = 0 \bmod p^{k+1}$ for all $J \in \mathbf{S}(n,r)$ with $|J| < n-k(p-1)$. Furthermore, the set of such S_J are linearly independent over $\mathbb{Z}/p$. The proof of the theorem now follows in much the same way as Proposition 4.5 of [4]. The idea of the proof is to first observe that the set $\{(\theta_{p-1})^{n-\deg I}\, \theta_I + \text{lots};\ I \in \mathbf{S}(n,r),\ \deg I \le n,\ |I| \ge n\}$ is orthogonal (in $\mathcal{M}(n,r) \otimes \mathbb{Z}/p$) to the vectors

$$\{S_J;\ J \in \mathbf{S}(n,r),\ |J| < n\}$$

and then proceed by induction.

COROLLARY 6.3. $|\mathcal{M}(n,r)/\mathcal{I}(n,r)| = \Pi p^{q(I)}$ _where_ $I \in \mathbf{S}(n,r)$ _with_ $\deg I \le n$ _and_ $q(I) = [(n - |I| + p-2)/(p-1)]$.

Theorems 6.1 and 6.2 together prove the main theorem.

7. FINAL REMARK

Theorems 2.1 and 2.r $(2 \le r \le p-1)$ have some very interesting consequences.

Suppose that $S_1, S_2, \ldots, S_k$ is a collection of unitary spheres of dimension $2n-1$, each with a free linear $\mathbb{Z}/p$ action. If the disjoint union $S_1 \cup S_2 \cup \ldots \cup S_k$ is freely $\mathbb{Z}/p$ bordant to zero then from the manifolds in Theorem 2.1, we may construct an explicit $\mathbb{Z}/p$ free manifold whose boundary is precisely $S_1 \cup S_2 \cup \ldots \cup S_k$. In particular we can construct explicitly a $\mathbb{Z}/p$ free manifold M^{2n} whose boundary is $p^{[n/(p-1)]}$ copies of any unitary sphere of dimension 2n with a free linear $\mathbb{Z}/p$ action.

Details are left to the reader as is the formulation of the analogous interpretation of Theorem 2.r $(2 \le r \le p-1)$.

REFERENCES

1. M. F. Atiyah and G. B. Segal, "Equivariant K-theory", University of Warwick Notes.

2. L. Illusie, "Nombres de Chern et groupes finis", Topology 7 (1968), 255-269.

3. C. Kosniowski, "The equivariant bordism ring of $\mathbb{Z}/p$ manifolds with isolated fixed points". Proceedings of a conference on Topology and its applications. (Newfoundland 1973). New York, Marcel Dekker 1975, pp. 157-165.

4. C. Kosniowski, "$\mathbb{Z}/p$ manifolds with isolated fixed points". Math. Z. 136 (1974), 179-191.

5. C. Kosniowski, "Generators of the $\mathbb{Z}/p$ bordism ring". Math. Z 149 (1976), 121-130.

6. C. Kosniowski, "Characteristic numbers of $\mathbb{Z}/p$ manifolds". Journal of the L.M.S.

UNIVERSITY OF NEWCASTLE UPON TYNE

GAPS IN THE RELATIVE DEGREE OF SYMMETRY

HSU-TUNG KU AND MEI-CHIN KU

1. INTRODUCTION

Suppose M^m is a connected differentiable m-manifold. We define the _relative degree of symmetry_ FN(M) of M as the supremum of the dimensions of the compact Lie groups which can act effectively and differentiably on M with non-empty fixed point set. A closely related concept is the _degree of symmetry_ N(M) of M defined by W. Y. Hsiang [5] which is defined as the maximum of the dimensions of the compact Lie groups acting effectively and differentiably on M. Of course we may consider N(M) as the maximum of the dimensions of the isometry groups of all possible Riemannian structures on M if M is compact. It is easy to verify that $FN(M^m) \leq (m-1)m/2$ if $m \geq 1$, and $N(M^m) \leq m + FN(M^m)$. Consequently, if we want to estimate the upper bound of $N(M^m)$, it is enough to estimate $FN(M^m)$. Moreover, the relative degree of symmetry can be used to study the dimensions of certain non-compact Lie transformation groups on connected differentiable manifolds. For instance, we have proved the following results in [11], where Theorem 1.2 is an easy consequence of Theorem 1.1.

THEOREM 1.1 [11]. _Let_ M^m _be a connected differentiable m-manifold_, $m \geq 19$. _Then precisely one of the following holds._

(i) $FN(M) \le \langle\alpha\rangle + \langle m-\alpha\rangle - m$ for all α such that $H^{\alpha}(M;K_p) \neq 0$,

where K_p denotes a field of characteristic $p \neq 2$, and $\langle n\rangle$ denotes $n(n+1)/2$ for a non-negative integer n.

(ii) M is diffeomorphic to CP^k $(m = 2k)$ and $FN(M) = \dim U(k)$.

THEOREM 1.2 [10, 11]. Let I(M) denote the isometry group of the connected Riemannian m-manifold M, $m \geq 19$. Then precisely one of the following holds.

(i) $\dim I(M) \leq \langle\alpha\rangle + \langle m-\alpha\rangle$ for all α such that $H^{\alpha}(M;K_p) \neq 0$.

(ii) $M = CP^k$ $(m = 2k)$, and $\dim I(M) = \dim SU(k+1)$.

It is proved in [13] that a general pattern of gaps exist in the degree of symmetry. In fact, $N(M^m)$ cannot fall into any of the following ranges if $m \geq 17$:

$$\langle m-k\rangle + \langle k\rangle < N(M^m) < \langle m-k+1\rangle, \quad k = 1,2,3,\ldots .$$

In this paper we shall show that similar gaps exist in FN(M). These results will be used to obtain the gaps in the dimensions of the isometry groups of the Riemannian manifolds. The main results are as follows:

THEOREM A. Let M be a connected differentiable m-manifold, $m \geq 17$. Then FN(M) cannot fall into any of the following ranges:

$$\langle m-k-1\rangle + \langle k-1\rangle < FN(M) < \langle m-k\rangle, \quad k = 1,2,\ldots,\psi(m),$$

where $\psi(m)$ is the largest integer j satisfying the inequality

$$\langle m-j-1\rangle + \langle j-1\rangle < \langle m-j\rangle .$$

THEOREM B. Let M be a connected differentiable m-manifold and k_i $(i = 0,1,\ldots,s+1)$ be any sequence of positive integers satisfying the following conditions:

$$k_o = m, \; k_{i+1} \leq \psi(k_i), \; 0 \leq i \leq s, \text{ and } k_s \geq 18.$$

Then FN(M) cannot fall into the following range:

$$\Sigma_{i=0}^{s-1} \langle k_i-k_{i+1}-1\rangle + \langle k_s-k_{s+1}-1\rangle + \langle k_{s+1}-1\rangle < FN(M)$$

$$< \Sigma_{i=0}^{s-1} \langle k_i-k_{i+1}-1\rangle + \langle k_s-k_{s+1}\rangle .$$

For a positive integer m, let $\Phi(m)$ denote the largest integer j satisfying the inequality

$$\langle m-j\rangle + \langle j\rangle < \langle m-j+1\rangle .$$

THEOREM C. *Let G be a closed subgroup of the isometry group I(M) of a connected Riemannian m-manifold M, and* k_i $(i = 0,1,\ldots,s+1)$ *be any sequence of positive integers satisfying the conditions;* $k_0 = m$, $k_{i+1} \leq \Phi(k_i)$, $0 \leq i \leq s$, *and* $k_s - k_{s+1} \geq 18$. *Then the dimension of G cannot fall into the following range:*

$$(1) \qquad \Sigma_{i=0}^{s-1} \langle k_i-k_{i+1}\rangle + \langle k_s-k_{s+1}\rangle + \langle k_{s+1}\rangle < \dim G$$

$$< \Sigma_{i=0}^{s-1} \langle k_i-k_{i+1}\rangle + \langle k_s-k_{s+1}+1\rangle .$$

Theorem C generalizes the further gaps theorem in [14].

2. PRELIMINARIES

Let G be a compact connected Lie group. Following Janich [8], we let m(G) denote the minimal dimension of the connected manifolds upon which G acts almost effectively, e.g.

$$m(Sp(n)) = 4n - 4 ,$$

$$m(SU(n)) = 2n - 2 ,$$

$$m(SO(n)) = n - 1 .$$

We shall always express a compact connected Lie group G in the following form, and call the G_i's the *normal factor* of G (or $\overline{G}$).

(2) $$G = (T^q \times G_1 \times G_2 \times \dots \times G_v)/N = \overline{G}/N ,$$

where T^q is a q-torus $(q \geq 0)$, each G_i is either simple, simply connected or isomorphic to $\mathrm{Spin}(4) \cong \mathrm{Spin}(3) \times \mathrm{Spin}(3)$, and there is at most one Spin(3), and N is a finite normal subgroup of $\overline{G}$.

LEMMA 2.1 [12, 13]. *If* $n_1 \geq n_2 \geq n_3 \geq 0$, *then*

(i) $\langle n_1 \rangle + \langle n_2 \rangle \leq \langle n_1 + n_2 \rangle$,

(ii) $\langle n_1 \rangle + \langle n_2 \rangle \leq \langle n_1 + n_3 \rangle + \langle n_1 - n_3 \rangle$.

LEMMA 2.2 [13]. *Let* G *be a compact connected Lie group acting almost effectively on a connected manifold* M, *and* t *denote the dimension of a principal orbit. If* G *has a decomposition of the form* (2), *then there exists integers* $t_i \geq m(G_i)$ *such that* $\dim G_i \leq \langle t_i \rangle$, $1 \leq i \leq v$, *and* $\Sigma_{i=1}^{v} t_i \leq t - q$.

COROLLARY 2.3. *Suppose the compact connected Lie group* G *acts almost effectively on a connected m-manifold* M. *Let the integers* t_j's *satisfy*

(i) $\dim G_i \leq \langle t_i \rangle$, $1 \leq i \leq v$, *and* $\Sigma_{i=1}^{v} t_i \leq m - q$,

(ii) $t_j \leq t_1 \leq \gamma \leq m$, $2 \leq j \leq v$, *for some integer* γ .

Then $\dim G \leq \langle \gamma \rangle + \langle m - \gamma \rangle$.

PROOF. Let $t_1 = \gamma - u$, $u \geq 0$. Then by the hypothesis (i), we have

$$\dim G \leq \langle \gamma - u \rangle + \Sigma_{j=2}^{v} \langle t_j \rangle + q ,$$

and

(3) $$\Sigma_{j=2}^{v} t_j + q \leq m - t_1 = m - \gamma + u .$$

We consider two cases:

(a) $\Sigma_{j=2}^{v} t_j + q \leq u$. Then by Lemma 2.1 (i)

$$\Sigma_{j=2}^{v}\langle t_j\rangle + q \le \langle\Sigma_{j=2}^{v} t_j + q\rangle \le \langle u\rangle .$$

It follows that

$$\dim G = \dim \overline{G} \le \langle\gamma-u\rangle + \langle u\rangle \le \langle\gamma\rangle \le \langle\gamma\rangle + \langle m-\gamma\rangle .$$

(b) $\Sigma_{j=2}^{v} t_j + q > u$. By repeated use of Lemma 2.1 (ii),

$$\langle\gamma-u\rangle + \Sigma_{j=2}^{v}\langle t_j\rangle + q \le \langle\gamma\rangle + \Sigma_{j=2}^{v}\langle\tilde{t}_j\rangle + \tilde{q} ,$$

where $0 \le \tilde{t}_j \le t_j$, $0 \le \tilde{q} \le q$, and

$$\Sigma_{j=2}^{v}\tilde{t}_j + \tilde{q} = \Sigma_{j=2}^{v} t_j + q - u .$$

Consequently,

$$\dim G \le \langle\gamma\rangle + \Sigma_{j=2}^{v}\langle\tilde{t}_j\rangle + \tilde{q} < \langle\gamma\rangle + \langle\Sigma_{j=2}^{v}\tilde{t}_j + \tilde{q}\rangle$$

$$= \langle\gamma\rangle + \langle\Sigma_{j=2}^{v} t_j + q - u\rangle$$

$$\le \langle\gamma\rangle + \langle m - \gamma\rangle \qquad \text{(from (3))}.$$

LEMMA 2.4 [6, 7, 12]. *Let G be a compact Lie group acting almost effectively on G/H and $\dim G > r \dim G/H$, $r > 13/4$. Then there exists a normal factor G_1 of G such that $\dim G_1 > rm(G_1)$, and G_1 is isomorphic to one of the following:*

(i) $\mathrm{Spin}(n)$, $n > 2r$,

(ii) $SU(n)$, $n > 2r - 1$,

(iii) $Sp(n)$, $n > 2r - 2$.

Moreover,

$$\dim G_1 + \dim N(H_1,G_1)/H_1 > r \dim G_1/H_1 ,$$

where $H_1 = G_1 \cap H$, H a principal isotropy subgroup of the $\overline{G}$-action on M, and $N(H_1,G_1)$ denotes the normalizer of H_1 in G_1.

LEMMA 2.5. *Let* $G = G_1 \times K$ *be a compact connected Lie group acting differentiably and effectively on a differentiable manifold* M. *Suppose* $\dim G_1 = N(G_1(x))$, *where* $G_1(x)$ *is a principal* G_1*-orbit in* M. *Then* K *acts almost effectively on the orbit space* M/G_1.

PROOF. Suppose H is a non-trivial connected subgroup of K which acts trivially on M/G_1. Then $G_1(y)$ is an invariant H-subspace for every y in M. Since the union of all principal G-orbits in M is open and dense in M, hence H must act non-trivially on at least one principal G_1-orbit, say $G_1(x)$. It follows that $N(G_1(x)) > \dim G_1$ which is a contradiction to the hypothesis.

3. GAPS IN FN(M)

MAIN LEMMA. *Let* G *be a compact connected Lie group acting effectively and differentiably on a connected differentiable m-manifold* M *with non-empty fixed point set*, $m \geq 20$. *If*

$$\dim G > m^2/4 - m/2 ,$$

then exactly one of the following holds.

(α) $M = CP^{m/2}$, $G = SU(m/2)$ *or* $U(m/2)$ *and* $FN(M) = \dim U(m/2)$.

(β) $M = S^m$ *or* R^m, $G = SU(m/2)$, $U(m/2)$ *or* $SO(m)$ *and* $FN(M) = \dim SO(m)$.

(γ) $M = RP^m$, $G = SO(m)$ *and* $FN(M) = \dim SO(m)$.

(δ) $\bar{G} = U((m-1)/2)$ *and* $\bar{G}$ *acts on* M *with* S^{m-2} *as principal orbit*.

(ϵ) $\bar{G} = \text{Spin}(n) \times K$, $n > m/2$, *and the identity component of the principal isotropy subgroup of the* Spin (n) *action on* M *is a standard imbedded* Spin (n-1) *in* Spin (n).

PROOF. Let r be the dimension of a principal G-orbit. Since the action of G on M has non-empty fixed point set, $r = m - 1$, or $r \leq m - 2$. We divide the proof into two cases.

Case I. $r = m - 1$. By [17], the orbit space M/G is homeomorphic to either [0,1] or [0,1) because the fixed point set is not empty. Let x be a fixed point. Then the group G acts orthogonally on a neighbourhood of x, and hence the principal orbit is of type S^{m-1}. It follows that G acts effectively and transitively on S^{m-1}, and G is one of the following groups [1, 16, 19]

(a) $G = Sp(m/4)$, $(Sp(m/4) \times S^1)/Z_2$ or $(Sp(m/4) \times Sp(1))/Z_2$,

(b) $G = SU(m/2)$ or $U(m/2)$.

(c) $G = SO(m)$.

Case (a) is impossible. If not, then $\dim G \leq \dim Sp(m/4) + 3 = m^2/8 + m/4 + 3 < m^2/4 - m/2$, which is an obvious contradiction. Hence case (a) is eliminated.

If $M/G \approx [0,1]$, then M is compact. According to [17], if H is a principal isotropy subgroup of G, and G(y) is the singular orbit, with $G(y) \neq x$, then $G_y/H \approx S^d$, $d \geq 0$, and G acts orthogonally on R^{d+1} and transitively on S^d. Moreover,

$$(4) \qquad M \approx (G \times_{G_y} D^{d+1}) \cup_{S^{m-1}} (G \times_G D^m) .$$

Now, if $G = SU(m/2)$, then $H = SU(m/2-1)$, and $G_y = SU(m/2)$ or $U(m/2-1)$. It follows that

$$M \approx (G \times_G D^m) \cup_{S^{m-1}} (G \times_G D^m) \approx S^m,$$

or

$$M \approx (G \times_{U(m/2-1)} D^2) \cup_{S^{m-1}} (G \times_G D^m) \approx CP^{m/2} .$$

Similarly, if $G = U(m/2)$, then $M \approx S^m$, or $M \approx CP^{m/2}$.

In the case (c), $G = SO(m)$ and $G_y = SO(m)$ or $N(SO(m-1), SO(m))$. Hence it follows from (4) that $M \approx S^m$, or $M \approx RP^m$.

If $M/G \approx [0,1)$, then $M \approx G \times_G R^m \approx R^m$ by [2, p.205] or [17]. Since $FN(M^m) \leq \langle m-1 \rangle$, we have proved that we can have the possibilities (α), (β) and (γ) except $FN(CP^{m/2}) = \dim U(m/2)$.

Case II. $r \leq m-2$. Let $r = m - 2 - u$, $u \geq 0$. Then

$$\dim G > m^2/4 - m/2 > (m/4 + u/4)r .$$

By making use of Lemma 2.4, G contains a normal factor G_1 which is isomorphic to one of the following groups.

(i) $Sp(n)$, $n > (m+u)/2 - 2$,

(ii) $SU(n)$, $n > (m+u)/2 - 1$,

(iii) $Spin(n)$, $n > (m+u)/2$.

(i) can easily be eliminated. If not, we have

$$2(m+u) - 12 < 4n - 4 = m(Sp(n)) \leq r = m - 2 - u ,$$

or $m < 10 - 3u$, which contradicts $m \geq 20$.

Suppose now that (ii) holds. Let H_1 be a principal isotropy subgroup of $G_1 = SU(n)$. Since we are assuming $m \geq 20$, it follows that

$$\dim G_1/H_1 = \dim SU(n)/H_1 \leq m-2 \leq (1/2)(n-1)^2.$$

Hence H_1 contains a subgroup which is conjugate to a standard imbedde $SU(n-k)$ [4]. We may assume that

$$SU(n-k) \leq H_1 \leq S(U(n-k) \times U(k))$$

because the identity component of $N(SU(n-k), SU(n))$ is $S(U(n-k) \times U(k)$

Therefore

$$\dim N(H_1,G_1)/H_1 \le \dim SU(k) + 1 ,$$

and

$$m > \dim G_1/H_1 \ge \dim SU(n)/S(U(n-k) \times U(k)) = 2nk - 2k^2.$$

By Lemma 2.4, we have

$$\begin{aligned}\dim SU(n) + \dim SU(k) + 1 &\ge \dim SU(n) + \dim N(H_1,G_1)/H_1 \\ &> (m/4 + u/4) \dim SU(n)/S(U(n-k) \times U(k)) \\ &> ((2nk - 2k^2)/4)(2nk - 2k^2) .\end{aligned}$$

This implies that there is an inequality

$$k^4 - 2nk^3 + (n^2 - 1)k^2 - n^2 + 1 < 0 ,$$

which is possible only when $k = 1$. Since the fixed point set is non empty, $H_1 = SU(n-1)$. Thus

$$m + u - 2 \le 2n - 1 = \dim G_1/H_1 \le m - 2 - u .$$

It follows that $u = 0$ and $2n - 1 = m - 2 = r$. Hence $\bar{G}$ acts transitively on S^{m-2}. But $\bar{G} \supset SU(n)$, hence $\bar{G} = SU((m-1)/2)$ or $U((m-1)/2)$ (cf. (b)). Note that

$$\dim SU((m-1)/2) = m^2/4 - m/2 - 3/4 < m^2/4 - m/2 .$$

Consequently, $\bar{G} = U((m-1)/2)$. This gives the possibility (δ).

We can show as in case (ii) by using Lemma 2.4 that if $G_1 = Spin(n)$, then $H_1 = Spin(n-1)$, where H_1 is the principal isotropy subgroup of the action of G_1. Since $m \ge 20$, $Spin(n-1)$ imbedded standardly in $Spin(n)$ by [4]. Hence we have possibly (ϵ) of the Main Lemma.

To show that $FN(CP^{m/2}) = \dim U(m/2)$, it is enough to show that the manifold $CP^{m/2}$ cannot satisfy (ϵ). Otherwise, we consider the projection $\pi: CP^{m/2} \to CP^{m/2}/Spin(n)$. For each $x \in CP^{m/2}/Spin(n)$, $\pi^{-1}(x)$ is S^{n-1}, RP^{n-1} or fixed point, all of which are acyclic over Q up to and including dimension n-2. It follows from the Vietoris-Begle mapping theorem that

$$\pi^*: H^2(CP^{m/2}/Spin(n);Q) \cong H^2(CP^{m/2};Q) .$$

Hence $\dim CP^{m/2}/Spin(n) = m$ which is clearly impossible.

REMARK 3.1. In the Main Lemma, if we assume that $\dim G > m^2/4 + m/2$, $m \geq 17$, and $\dim G \neq \langle n-1 \rangle$, then only the possibility (ϵ) survives with $n > m/2 + 2$.

Let us recall the definition of $\psi(m)$, m a positive integer. It is defined as the largest integer j satisfying the inequality

$$\langle m - j - 1 \rangle + \langle j - 1 \rangle < \langle m - j \rangle .$$

It is easy to show that

(5) $$m > \langle j - 1 \rangle + j , \quad j = 1,2,\ldots,\psi(m) ,$$

and

(6) $$\langle m - j - 1 \rangle \geq \langle m - \psi(m) - 1 \rangle > m^2/4 + m/2 ,$$

for $j = 1,2,\ldots,\psi(m)$ if $m \geq 17$.

THEOREM 3.2. *Let G be a compact connected Lie group acting effectively and differentiably on a connected differentiable m-manifold with non-empty fixed point set, $m \geq 17$. Then the dimensi of G cannot fall into any of the following ranges:*

(7) $$\langle m - k - 1 \rangle + \langle k - 1 \rangle < \dim G < \langle m - k \rangle, \quad k = 1,2,\ldots,\psi$$

PROOF. If, on the contrary, then dim G satisfies (7). It follows from (6) that

$$\dim G > \langle m - \psi(m) - 1 \rangle > m^2/4 + m/2.$$

Hence we must have possibility (ϵ) of the Main Lemma by Remark 3.1.

Thus G contains a normal factor $G_1 = \mathrm{Spin}(n)$, $n > m/2 + 2$. Let $G(x)$ be a principal G-orbit of dimension r. By Main Lemma, $r \le m - 2$. Now

$$\langle n - 1\rangle = \dim G_1 \le \dim G < \langle m - k\rangle .$$

Consequently, if we let $t_1 = n - 1$, then $t_1 \le m - k - 1$. As $t_1 > m/2 + 1$, it follows from Lemma 2.2 that $t_j \le t_1$ for $2 \le j \le v$. Apply Corollary 2.3 to the action of G on $G(x)$, we get

$$\dim G \le \langle m - k - 1\rangle + \langle r - (m - k - 1)\rangle \le \langle m - k - 1\rangle + \langle k - 1\rangle,$$

which is a contradiction. This completes the proof of the theorem.

Now let G be a compact connected Lie group acting effectively on a topological manifold M with principal orbit of dimension t. Then $\bar{G}$ acts transitively and almost effectively on a principal G-orbit, say $\tilde{M}$. The principal T^q-isotropy subgroup is trivial. Hence the dimension of $M_0 = \tilde{M}/T^q$ is equal to $t - q$. The group $G_o = \bar{G}/T^q$ acts almost effectively and transitively on the compact manifold M_0. Let V be a direct product of G_i's such that V acts almost freely on M_0 and V is of maximal dimension. Following the proof of [13, Theorem 1] the group $\bar{G}$ can be decomposed as

(8) $$\bar{G} = T^q \times G_o = T^q \times H \times K \times V,$$

where H,K and V are each direct products of G_i's (see (2) for notation) and (9) K and V act almost freely on M_0 with $\dim K \le \dim V = w$, $H = G_{j_1} \times \dots \times G_{j_k}$ acts almost effectively on $M_1 = M_0/V$, $\dim G_{j_i} \le \langle t_{j_i}\rangle$, $1 \le i \le k$, and $\Sigma_{i=1}^k t_{j_i} = \dim M_0 - \dim V = t - q - w$.

THEOREM 3.3. _Let_ G _be a compact connected Lie group acting effectively and differentiably on a differentiable m-manifold_ M _with_ *non-empty fixed point set. Suppose* k_i $(i = 0,1,\ldots,s+1)$ *is any sequence of positive integers satisfying the following condition:*

$$(10) \qquad k_0 = m,\ k_{i+1} \le \psi(k_i),\ 0 \le 1 \le s,\ \text{and}\ k_s \ge 18.$$

Then $\dim G$ *cannot fall into the following range:*

$$(11) \qquad \Sigma_{i=0}^{s-1} \langle k_i - k_{i+1} - 1\rangle + \langle k_s - k_{s+1} - 1\rangle + \langle k_{s+1} - 1\rangle$$

$$< \dim G < \Sigma_{i=0}^{s-1} \langle k_i - k_{i+1} - 1\rangle + \langle k_s - k_{s+1}\rangle .$$

PROOF. The proof will be by induction on s. The assertion is certainly true when $s = 0$ which is precisely Theorem 3.2. If $s > 0$, then we may assume by induction that the assertion is true for $s - 1$. Suppose now that the dimension of G satisfies (11). Then $\dim G > m^2/4 + m/2$ by (6). It follows from Remark 3.1 that $\overline{G}$ contains a normal factor $G_1 = \mathrm{Spin}(n_1)$, $n_1 > m/2 + 2$, and $t_1 = n_1 - 1$. Suppose $\dim G_1 \ge \langle k_0 - k_1\rangle$. Then we have

$$\Sigma_{i=0}^{s-1}\langle k_i-k_{i+1}-1\rangle + \langle k_s-k_{s+1}\rangle$$

$$\le \langle k_0-k_1-1\rangle + \langle \Sigma_{i=1}^{s-1}(k_i-k_{i+1}-1) + k_s-k_{s+1}\rangle$$

$$= \langle k_0-k_1-1\rangle + \langle k_1-k_{s+1}-s+1\rangle$$

$$\le \langle k_0-k_1-1\rangle + \langle k_1-1\rangle \le \langle k_0-k_1\rangle \ (\text{since } k_1 \le \psi(k_0))$$

$$\le \dim G_1 \le \dim G,$$

which contradicts (11). Hence $\dim G_1 \le \langle k_0-k_1-1\rangle$, and $t_1 \le k_0 - k_1 -$

Now we proceed to show that $t_1 = k_0-k_1-1$. If not, $t_1 = k_0 - k_1 - u$, $u \ge 2$. Since $t_1 > k_0/2 + 1$, hence $2 \le u < k_0/2 - k_1$ Now we shall adapt the notation of (8) and (9). Since V is either a product of simple groups or identity, $\dim V = w \ge 3$, or 0. Clearly

G_1 must be a factor of H. We may assume that $G_1 = G_{j_1}$, and denote j_i simply by i, $1 \le i \le k$. Thus by (9),

$$\Sigma_{i=2}^{k} t_i = t - q - w - t_1 \le k_0 - 2 - q - w - t_1$$
$$= k_1 + u - q - w - 2.$$

We shall show that

(12) $$\langle k_0 - k_1 - u\rangle + \langle k_1 + u\rangle \le \langle k_0 - k_1 - 1\rangle + \langle k_1 - 1 - \psi(k_1)\rangle$$

if $2 \le u < k_0/2 - k_1 - 1$. To verify (12), it suffices to prove the following inequality

$$u(2k_0 - 4k_1 - 2u) - 2k_0 + \psi(k_1)^2 - 2k_1\psi(k_1) + \psi(k_1) \ge 0.$$

Let $f(u) = u(2k_0 - 4k_1 - 2u) - 2k_0 + \psi(k_1)^2 - 2\psi(k_1) + \psi(k_1)$. Then $f'(u) > 0$ if $u < k_0/2 - k_1 - 1$. Hence f(u) is increasing for $2 \le u < k_0/2 - k_1 - 1$. By making use of the hypotheses $\psi(k_0) \ge k_1$, $k_0 \ge \langle\psi(k_0) - 1\rangle + \psi(k_0) + 1$, and $k_s \ge 18$. We see easily that $f(2) \ge 0$. This proves (12). It follows that

$$\dim G \le \langle k_0 - k_1 - u\rangle + \langle\Sigma_{i=2}^{k} t_i\rangle + 2w + q$$
$$\le \langle k_0 - k_1 - u\rangle + \langle k_1 + u - q - w - 2\rangle + 2w + q$$
$$\le \langle k_0 - k_1 - u\rangle + \langle k_1 + u\rangle - \langle q + w + 2\rangle + 2w + q$$
$$< \langle k_0 - k_1 - u\rangle + \langle k_1 + u\rangle$$
$$\le \langle k_0 - k_1 - 1\rangle + \langle k_1 - 1 - \psi(k_1)\rangle \quad \text{(by (12))}$$
$$\le \langle k_0 - k_1 - 1\rangle + \langle k_1 - k_2 - 1\rangle \quad (\text{since } k_2 \le \psi(k_1))$$

which is a contradiction. This proves that $t_1 = k_0 - k_1 - 1$. Since the principal orbit type of G_1-action is either S^{n_1-1} or RP^{n_1-1}, it follows from Lemma 2.5 that the group $K = \overline{G}/G_1$ acts almost effectively on the orbit space M/G_1, and hence K acts almost effectively on the fixed point set $F(G_1, M) = \partial(M/G_1)$, the boundary of M/G_1 which is of dimension k_1. However

$$F(K, F(G_1, M)) = F(\overline{G}, M) \neq \emptyset ,$$

and

$$\Sigma_{i=1}^{s-1}\langle k_i - k_{i+1} - 1\rangle + \langle k_s - k_{s+1} - 1\rangle + \langle k_{s+1} - 1\rangle < \dim K$$

$$< \Sigma_{i=1}^{s-1}\langle k_i - k_{i+1} - 1\rangle + \langle k_s - k_{s+1}\rangle .$$

This is impossible by the inductive hypothesis. This concludes the proof of the theorem.

Theorems A and B are immediate consequences of Theorem 3.2 and Theorem 3.3.

4. GAPS IN THE DIMENSIONS OF THE ISOMETRY GROUPS OF THE RIEMANNIAN MANIFOLDS.

In this section we shall apply the results proved in the previous section to obtain the gaps in the dimensions of the isometry groups of the Riemannian manifolds.

THEOREM 4.1 [15]. _Let_ G _be a closed subgroup of the isometry group_ I(M) _of a connected Riemannian m-manifold_ M, $m \geq 17$. _Then the dimension of_ G _cannot fall into any of the following ranges:_

(13) $\langle m-k\rangle + \langle k\rangle < \dim G < \langle m-k+1\rangle, \quad k = 1, 2, \ldots, \Phi(m)$.

PROOF. If, contrary to the theorem, then there is a closed subgroup G of I(M) satisfying (13). Let G(x) be a principal G-orbit, and $H = G_x^o$. Then

$$\dim G = r + \dim H ,$$

where $r = \dim G(x)$. The group G acts effectively on $G(x)$ (cf. [3,9]), hence

$$\langle m-k\rangle < \dim G \leq \langle r\rangle ,$$

or $m-k+1 \leq r$. It follows from (13) that

$$\langle m-k-1\rangle + \langle k-1\rangle = \langle m-k\rangle + \langle k\rangle - m \leq \langle m-k\rangle + \langle k\rangle - r$$

$$< \dim H < \langle m-k+1\rangle - r \leq \langle m-k+1\rangle - (m-k+1) = \langle m-k\rangle .$$

This contradicts Theorem 3.2 because the fixed point set $F(H,M) \neq \emptyset$.

A different proof can be found in [15].

PROOF OF THEOREM C. We shall prove the theorem by induction on s. It is true for $s = 0$ by Theorem 4.1. Suppose dim G falls into the range (1); i.e.,

$$(1)\qquad \Sigma_{i=0}^{s-1}\langle k_i-k_{i+1}\rangle + \langle k_s-k_{s+1}\rangle + \langle k_{s+1}\rangle < \dim G$$

$$< \Sigma_{i=0}^{s-1}\langle k_i-k_{i+1}\rangle + \langle k_s-k_{s+1}+1\rangle .$$

Let $G(x)$ be a principal orbit of dimension r. Then

$$\dim G = r + \dim H ,$$

where $H = G_x^o$. Thus

$$\Sigma_{i=0}^{s-1}\langle k_i-k_{i+1}-1\rangle + \langle k_s-k_{s+1}-1\rangle + \langle k_{s+1}-1\rangle$$

$$\leq \Sigma_{i=0}^{s-1}\langle k_i-k_{i+1}\rangle + \langle k_s-k_{s+1}\rangle + \langle k_{s+1}\rangle - r$$

$$< \dim H < \Sigma_{i=0}^{s-1}\langle k_i-k_{i+1}-1\rangle + \langle k_s-k_{s+1}\rangle + (k_0-k_{s+1}+1-r).$$

By Theorem 3.3, $r \leq k_0-k_{s+1}$. Let $W^{k_0-k_{s+1}} = G(x) \times S^{k_0-k_{s+1}-r}$. Now let us consider the natural effective action of G on W. Let

$$\overline{k}_i = k_i-k_{s+1}, \quad i = 0,1,\ldots,s.$$

Then $\bar{k}_i$ $(i = 0,1,\ldots,s)$ satisfies

$$\bar{k}_{i+1} \leq \Phi(\bar{k}_i), \quad 0 \leq i \leq s-1, \text{ and } \bar{k}_s \geq 18 .$$

By induction, dim G cannot fall into the following range.

$$\Sigma_{i=0}^{s-2}\langle \bar{k}_i - \bar{k}_{i+1}\rangle + \langle k_{s-1} - \bar{k}_s\rangle + \langle \bar{k}_s\rangle < \dim G$$

$$< \Sigma_{i=0}^{s-2}\langle \bar{k}_i - \bar{k}_{i+1}\rangle + \langle \bar{k}_{s-1} - \bar{k}_s + 1\rangle ,$$

that is, dim G cannot fall into the following range

$$\Sigma_{i=0}^{s-2}\langle k_i - k_{i+1}\rangle + \langle k_{s-1} - k_s\rangle + \langle k_s - k_{s+1}\rangle < \dim G$$

$$< \Sigma_{i=0}^{s-2}\langle k_i - k_{i+1}\rangle + \langle k_{s-1} - k_s + 1\rangle .$$

However,

$$\Sigma_{i=0}^{s-1}\langle k_i - k_{i+1}\rangle + \langle k_s - k_{s+1}\rangle + \langle k_{s+1}\rangle < \dim G .$$

It follows that

$$\dim G \geq \Sigma_{i=0}^{s-2}\langle k_i - k_{i+1}\rangle + \langle k_{s-1} - k_s + 1\rangle$$

$$\geq \Sigma_{i=0}^{s-2}\langle k_i - k_{i+1}\rangle + \langle k_{s-1} - k_s\rangle + \langle k_s\rangle \quad (\text{since } k_s \leq \Phi(k_{s-1}))$$

$$\geq \Sigma_{i=0}^{s-1}\langle k_i - k_{i+1}\rangle + \langle k_s - k_{s+1} + 1\rangle.$$

This contradicts (1).

References

1. A. Borel. Some remarks about Lie groups transitive on spheres and tori. Bull. Amer. Math. Soc. 55 (1949), 580-586.
2. G.E. Bredon. Introduction to compact transformation groups. Academic Press, New York and London, 1972.
3. R. Hermann. Differential geometry and the calculus of variations. Academic Press, New York and London, 1968.
4. W.C. Hsiang and W.Y. Hsiang. Differentiable actions of compact connected classical groups, I. Amer. J. Math. 89 (1967), 705-786.
5. W.Y. Hsiang. On the bound of the dimensions of the isometry groups of all possible Riemannian metrics on an exotic sphere. Ann. of Math. 85 (1967), 351-358.
6. W.Y. Hsiang. The natural metric on SO(n)/SO(n-2) is the most symmetric metric. Bull. Amer. Math. Soc. 73 (1967), 55-58.
7. W.Y. Hsiang. On the degree of symmetry and the structure of highly symmetric manifolds. Tankang J. of Math. (Taipei) 2 (1971), 1-22.
8. K. Janich. Differenzierbare G-Mannighfaltigkeiten. Lecture Notes in Math., No. 59, Springer-Verlag, Berlin and New York, 1968.
9. S. Kobayashi and T. Nagano. Riemannian manifolds with abundant isometries. Diff. Geometry in honour of K. Yano, 195-220, Kinokuniya, Tokyo, 1972.
10. H.T. Ku and M.C. Ku. Dimension of a differentiable transformation group. Bull. Inst. Math., Academia Sinica (Taipei) 3 (1975), 29-41.

11. H.T. Ku and M.C. Ku. Degree of symmetry of manifolds. Seminar Notes, Univ. of Mass., Amherst, 1976.

12. H.T. Ku, L.N. Mann, J.L. Sicks and J.C. Su. Degree of symmetry of a product manifold. Trans. Amer. Math. Soc. 146 (1969), 133-149.

13. L.N. Mann. Gaps in the dimension of transformation groups. Illinois J. Math. 10 (1966), 532-546.

14. L.N. Mann. Further gaps in the dimensions of transformation groups. Illinois J. Math. 13 (1969), 740-756.

15. L.N. Mann. Gaps in the dimensions of isometry groups of Riemannian manifolds. J. of Diff. Geometry (to appear).

16. D. Montgomery and H. Samelson. Transformation groups on spheres. Ann. of Math. 44 (1943), 454-470.

17. P.S. Mostert. On a compact Lie group acting on a manifold. Ann. of Math. 65 (1957), 447-455.

18. S.B. Myers and N.E. Steenord. The group of isometries of a Riemannian manifold. Ann. of Math. 40 (1939), 400-416.

19. J. Poncet. Groupes de Lie compacts de transformations de l'espace euclidean et les sphères comme espaces homogènes. Comment. Math. Helv. 33 (1959), 109-120.

UNIVERSITY OF MASSACHUSETTS
AMHERST, MA 01002
U.S.A.

CHARACTERS DO NOT LIE

ARUNAS LIULEVICIUS

Let G be a compact topological group, $\alpha : G \to U(n+1)$ a representation of G and $P(\alpha)$ the corresponding G-structure on complex projective n-space $CP^n = U(n+1)/U(1)\times U(n)$. We investigate the G-homotopy types of these linear G-actions $P(\alpha)$. Our main result is:

THEOREM 1. *If G is a compact group*, $\alpha, \beta : G \to U(n+1)$ *representations of G and* $f : P(\alpha) \to P(\beta)$ *is a G-map which is a homotopy equivalence then there exists a homomorphism* $\chi : G \to S^1$ *such that either* β *or its conjugate* $\bar{\beta}$ *is similar to* $\chi\alpha$.

An immediate result of the theorem is that group characters do not lie: the G-homotopy type of $P(\alpha)$ can be read off the character table of G. For example, if $\alpha, \beta : G \to U(n+1)$ are representations and $|\text{Trace } \alpha(g)| \neq |\text{Trace } \beta(g)|$ for some $g \in G$ then $P(\alpha)$ is not G-homotopy equivalent to $P(\beta)$.

The reader suspects that the correct statement of Theorem 1 should be in terms of homomorphisms into the collineation group $PU(n+1) = U(n+1)/S^1$ of CP^n. This is the content of

COROLLARY 2. *If G is a compact group and* $\alpha, \beta : G \to PU(n+1)$ *are two homomorphisms then* (CP^n, α) *is G-homotopy equivalent to* (CP^n, β) *if and only if* β *or its complex conjugate* $\bar{\beta}$ *is similar to* α.

REMARK. The map $f : P(\alpha) \to P(\beta)$ in Theorem 1 is not assumed to be G-homotopy equivalence, that is although f has a homotopy inverse $f' : P(\beta) \to P(\alpha)$, we are not assuming that an f' can be found which is a G-map. Petrie [3] exhibits a G-space Y and a G-map $h : Y \to P(\alpha)$ which is a homotopy equivalence such that the induced map in equivariant K-theory $h^! : K_G(P(\alpha)) \to K_G(Y)$ is not an isomorphism (that is h is not a G-homotopy equivalence). The next theorem shows that linear G-actions do not admit this type of pathology.

Our main tool in the proof of Theorem 1 is G-equivariant K-theory [1], [4]. Let $\rho : G \to U(n+1)$ be a representation and $s = s(\rho) : S(\rho) \to P(\rho)$ denote the projection from the unit sphere in C^{n+1} to CP^n. The projection s is G-equivariant and defines a G line bundle which we still call s. It is well known that

$$K_G(P(\rho)) = \Gamma[s]/(c(\rho))$$

where $\Gamma = R(G)$ is the complex representation ring of G and $c(\rho)$ is the characteristic polynomial of ρ:

$$c(\rho) = \sum_{i=0}^{n+1} (-1)^i (\Lambda^i \rho) s^{n+1-i},$$

where $\Lambda^i \rho$ is the i-th exterior power of the representation ρ (see [4], pp.142-3).

Suppose $f : P(\alpha) \to P(\beta)$ is a G-map with $f^* =$ identity on $H^*(CP^n;Z)$. Our next result shows that the induced map

$$f^! : K_G(P(\beta)) \to K_G(P(\alpha))$$

must have a very special form.

THEOREM 3. Let $f : P(\alpha) \to P(\beta)$ be a G-map with $f^* =$ identity, $s = s(\alpha)$, $t = s(\beta)$, then there exists a homomorphism

$\chi : G \to S^1$ _such that_ $f^! t = \chi s$.

Denote by $\mathrm{Pic}_G(X)$ the group of isomorphism classes of complex G line bundles over the G-space X with tensor product as group operation. Let $c : X \to P$ be the collapsing map onto the point P and $i : E \to G$ the inclusion of the identity sub-group E into G.

THEOREM 4. _If X is a path-connected G-space with_ $H^1(X;Z) = 0$ _then the following sequence is exact_

$$\mathrm{Pic}_G(P) \xrightarrow{c^!} \mathrm{Pic}_G(X) \xrightarrow{i^!} \mathrm{Pic}_E(X).$$

The paper is organised as follows:

§1. Proof of Theorem 4;

§2. Theorem 4 implies Theorem 3;

§3. Theorem 3 implies Theorem 1;

§4. Theorem 1 implies Corollary 2.

This paper is the result of numerous conversations and consultations with Ted Petrie. The proof of an earlier version of Theorem 4 (see [2]) is due to Petrie. Johan Dupont showed how to remove the hypothesis that G has a fixed point on X in that early version of Theorem 4 and Graeme Segal explained how [5] and [6] would simplify the argument. Thanks also go to Frank Adams, Peter Landrock and Jørgen Tornehave for their helpful comments and the Matematisk Institut, Aarhus Universitet for its hospitality.

1. PROOF OF THEOREM 4.

We shall use the cohomology groups $H^i_G(A)$ of Segal [5], where A is a locally contractible abelian group in the category of compactly generated Hausdorff G-spaces. A sequence of G-homomorphisms

$$A' \xrightarrow{i} A \xrightarrow{p} A''$$

is said to be a short exact sequence if i is an embedding onto a closed subgroup, p induces a topological isomorphism of A/A' with A'', and the fibration p is locally trivial topologically. A short exact sequence of groups induces a long exact sequence of cohomology groups. The group $H^1_G(A)$ is Hom(G,A), the group of all crossed homomorphisms of G into A modulo principal crossed homomorphisms (in particular if the action of G on A is trivial then it is the group of all homomorphisms). If A is contractible, then $H^i_G(A)$ can be calculated from the familiar complex of continuous cochains on G with values in A. In particular if A is a vector space over the field of real numbers R then $H^i_G(A) = 0$ for $i > 0$, because the complex of continuous cochains has a contracting chain homotopy (obtained by using the Haar integral on G). If X is a G-space we denote by Map(X,A) the set of all continuous functions $f : X \to A$ with the compact-open topology and G-action $(g.f)(x) = g.f(g^{-1}.x)$. Let P be a one point space and $c : X \to P$ the collapsing map, then the induced map $\mathrm{Map}(P,A) \to \mathrm{Map}(X,A)$ is the embedding of constant maps if $X \neq \emptyset$. We have to study the kernel of $i^! : \mathrm{Pic}_G(X) \to \mathrm{Pic}_E(X)$ where $i : E \to G$ is the inclusion of the trivial group. A G-structure on the trivial S^1-bundle $\pi_1 : X \times S^1 \to X$ corresponds to a crossed homomorphism $G \to \mathrm{Map}(X,S^1)$ and an isomorphism of such G-structures corresponds to a principal crossed homomorphism (here S^1 has trivial

G-action), so it is immediate that Ker $i^!$ = $H^1_G(\mathrm{Map}(X,S^1))$. To prove the theorem we have to show that the homomorphism $c: S^1 \to \mathrm{Map}(X,S^1)$ given by inclusion of constant functions induces an isomorphism of $H^1_G(\ )$. Consider the short exact sequence

$$Z \to R \overset{\exp}{\to} S^1.$$

It induces the short exact sequence

$$\mathrm{Map}(X,Z) \to \mathrm{Map}(X,R) \to \mathrm{Map}(X,S^1)$$

(the last map is onto since $H^1(X;Z) = 0$). Notice that Map(X,Z)=Z since X is connected. We have a map of the first short exact sequence into the second by inclusion of constant functions, hence a commutative diagram of cohomology groups

$$\begin{array}{ccccccc} H^1_G(R) & \to & H^1_G(S^1) & \overset{\delta}{\to} & H^2_G(Z) & \to & H^2_G(R) \\ \downarrow & & \downarrow c_* & & \downarrow = & & \downarrow \\ H^1_G(\mathrm{Map}(X,R)) & \to & H^1_G(\mathrm{Map}(X,S^1)) & \overset{\delta'}{\to} & H^2_G(Z) & \to & H^2_G(\mathrm{Map}(X,R)). \end{array}$$

Since the extreme terms vanish in both sequences (coefficients are vector spaces over R), δ and δ' are isomorphisms, so c_* is an isomorphism, and Theorem 4 is proved.

REMARK. Segal [6] proves that $\mathrm{Pic}_G(X) = H^2(EG\times_G X;Z)$, where $G \to EG \to BG$ is the classifying fibration. The sequence of Theorem 4 corresponds to the exact sequence

$$H^2(BG;Z) \overset{\pi^*}{\to} H^2(EG\times_G X;Z) \overset{i^*}{\to} H^2(X;Z)$$

where $X \overset{i}{\to} EG\times_G(X) \overset{\pi}{\to} BG$ is the fibration over BG with fiber X given by the Borel construction.

Since $\mathrm{Pic}_G(P) = \mathrm{Hom}(G,S^1)$, if $x \in$ Ker $i^!$ then $x = c^!(\chi)$ for a suitable homomorphism $\chi : G \to S^1$. In case X has a fixed point x_o (the hypothesis of the earlier version of the theorem in [2]) this homomorphism is easy to exhibit: it is defined by $g.(x_o,t) = (x_o,\chi(g)t)$, where we are looking at the G-structure on the trivial S^1-bundle $\pi_1 : X\times S^1 \to X$.

2. THEOREM 4 IMPLIES THEOREM 3.

Consider a homotopy equivalence $f : P(\alpha) \to P(\beta)$ with $f^*u = u$ for $u \in H^2(CP^n;Z)$, that is: $i^! f^! t = i^! s$, where $i : E \to G$ is the inclusion of the trivial group into G. We now apply Theorem 4: there exists a homomorphism $\chi : G \to S^1$ such that $s^{-1} f^! t = c^!(\chi) = \chi.1$, or $f^! t = \chi s$ as claimed.

3. THEOREM 3 IMPLIES THEOREM 1.

Let $\beta : G \to U(n+1)$ be a representation. Consider the complex conjugate $\bar{\beta}$ of β and the G-homotopy equivalence $\tilde{c} : S(\beta) \to S(\bar{\beta})$ defined by $\tilde{c}(z_o,\ldots,z_n) = (\bar{z}_o,\ldots,\bar{z}_n)$ - it induces a G-homotopy equivalence $c : P(\beta) \to P(\bar{\beta})$ and the following diagram commutes:

$$\begin{array}{ccc} S^1 & \xrightarrow{r} & S^1 \\ \downarrow & & \downarrow \\ S(\beta) & \xrightarrow{\tilde{c}} & S(\bar{\beta}) \\ \downarrow s & & \downarrow \bar{s} \\ P(\beta) & \xrightarrow{c} & P(\bar{\beta}) \end{array}$$

where $r(X) = x^{-1} = \bar{x}$, $s = s(\beta)$, $\bar{s} = s(\beta)$. That is

$$c^! : K_G(P(\bar{\beta})) \to K_G(P(\beta))$$

is given by $c^!\bar{s} = s^{-1}$ and

$$c^* : H^*(CP^n;Z) \to H^*(CP^n;Z)$$

is given by $c^*u = -u$, where $u \in H^2(CP^n;Z)$.

This example has two consequences: first, it shows that if β or $\bar{\beta}$ are similar to $\chi\alpha$, where $\chi : G \to S^1$ is a homomorphism, then $P(\alpha)$ is homotopy equivalent to $P(\beta)$; second, it shows that if $f : P(\alpha) \to P(\beta)$ is a homotopy equivalence then we may assume that $f^*u = u$ by replacing f by cf and β by $\bar{\beta}$ if necessary.

We let $t = s(\beta)$, $s = s(\alpha)$ and assume that f* is the identity map. According to Theorem 3 there exists a homomorphism $\chi : G \to S^1$ such that $f^! t = \chi s$. Recall that t satisfies the equation

$$t^{n+1} = \beta t^n - (\Lambda^2\beta)t^{n-1} + \dots + (-1)^n \Lambda^{n+1}\beta,$$

hence applying $f^!$ we have

$$\chi^{n+1}s^{n+1} = \beta\chi^n s^n - (\Lambda^2\beta)\chi^{n-1}s^{n-1} + \dots + (-1)^n \Lambda^{n+1}\beta,$$

and multiplying both sides by χ^{-n-1} we obtain

$$s^{n+1} = \beta\chi^{-1}s^n - (\Lambda^2\beta)\chi^{-2}s^{n-1} + \dots + (-1)^n(\Lambda^{n+1}\beta)\chi^{-n-1},$$

but

$$s^{n+1} = \alpha s^n - (\Lambda^2\alpha)s^{n-1} + \dots + (-1)^n \Lambda^{n+1}\alpha$$

and $K_G(P(\alpha))$ is a free R(G)-module on $\{s^n, s^{n-1}, \dots, 1\}$, so the coefficients of s^n are equal: $\alpha = \beta\chi^{-1}$, or $\beta = \chi\alpha$ as we wanted to prove.

4. THEOREM 1 IMPLIES COROLLARY 2

Suppose $\alpha, \beta : G \to PU(n+1)$ are representations. Let $p : U(n+1) \to PU(n+1)$ be the projection map and define the compact group $\tilde{G}$ as the pullback of $p \times p$ under the map (α, β), that is we have a commutative diagram of group homomorphisms

$$\begin{array}{ccc} \tilde{G} & \xrightarrow{(\tilde{\alpha},\tilde{\beta})} & U(n+1)\times U(n+1) \\ \downarrow{\scriptstyle \pi} & & \downarrow{\scriptstyle p\times p} \\ G & \xrightarrow{(\alpha,\beta)} & PU(n+1)\times PU(n+1). \end{array}$$

If $f : (CP^n,\alpha) \to (CP^n,\beta)$ is a G-homotopy equivalence, then since $\alpha\pi = p\tilde{\alpha}$ and $\beta\pi = p\tilde{\beta}$, $f : P(\tilde{\alpha}) \to P(\tilde{\beta})$ is a $\tilde{G}$-homotopy equivalence, so according to Theorem 1 either $\tilde{\beta}$ or its complex conjugate is similar to $\chi\tilde{\alpha}$ for a suitable homomorphism $\chi : \tilde{G} \to S^1$. Since $p(\chi\tilde{\alpha}) = p\tilde{\alpha} = \alpha$, Corollary 2 follows.

REFERENCES

1. M.F. Atiyah and G.B. Segal, Lectures on equivariant K-Theory. Mimeographed notes, Oxford 1965.
2. A. Liulevicius, Homotopy types of linear G-actions on complex projective spaces, Aarhus Universitet, Matematisk Institut Preprint Series 1975/76 No. 14.
3. T. Petrie, A setting for smooth S^1 actions with applications to real algebraic actions on $P(C^{4n})$, Topology 13 (1974), 363-
4. G. Segal, Equivariant K-theory, Publ. Math. I.H.E.S. 34 (1968 129-151.
5. G. Segal, Cohomology of topological groups, Istituto Naziona di Alta Matematica Symposia Mathematica 4 (1970), 377-387.
6. G. Segal, Cohomology of topological groups (to appear).

THE UNIVERSITY OF CHICAGO,
CHICAGO, IL. 60637
U.S.A.

ACTIONS OF $Z/2n$ ON S^3

GERHARD X. RITTER

ABSTRACT

This paper is devoted to classifying the actions of the cyclic group $Z/2n$ on the 3-sphere S^3. In particular, we show that if $h \in Z/2n$ is a generator, then h is equivalent to a standard rotation of S^3 if and only if h is a free action or has an almost tame fixed point set.

1. INTRODUCTION

The object of this paper is to classify all actions of the cyclic groups $Z/2n$ on the 3-sphere S^3. In section 2 we classify the non-free actions and in section 3 the free actions of $Z/2n$. In 1970, F. Waldhausen [19] has proven that every even periodic P.L. homeomorphism of S^3 with 1-dimensional fixed point set is topologically equivalent to a standard rotation of S^3. We shall extend Waldhausen's result by omitting the P.L. hypothesis and showing that every even periodic homeomorphism of S^3 with non-empty fixed point set is topologically a standard rotation if and only if the set of fixed points is almost tame. In view of R. H. Bing's work [1] and [2], this is the strongest possible generalization of Waldhausen's result. This also settles the Smith conjecture [3] for homeomorphisms of period 2n whose fixed point sets are almost tame knots.

The problem of characterizing free cyclic actions on S^3 remains largely unsolved. Thus far only free actions of Z/2 [6], Z/4 [11], and Z/8 [12] have been classified. Recently, this author [13] extended the list to all free actions of Z/2n, n = 1,2,... In this note we show how this last result can be generalized without great difficulty to free actions of Z/2n on S^3.

2. HOMEOMORPHISMS WITH ALMOST TAME FIXED POINT SET.

We shall denote the set of fixed points of a map h by Fix(h). Recall that a set S is _almost tame_ in a manifold M if and only if S is locally tame in M except possibly at a discrete set of points.

THEOREM 1. _Let_ h _be a periodic homeomorphism of_ S^3 _of period_ 2n. _If_ $\text{Fix}(h) \neq \emptyset$, _then_ h _is topologically equivalent to a standard rotation or reflection if and only if_ Fix(h) _is almost tame_.

The proof of Theorem 1 follows readily from a theorem of Kwun and Tollefson [5] and a powerful result recently obtained by Moise [8].

LEMMA 1 (Kwun and Tollefson [5]). _An involution_ h _of a 3-manifold is_ P.L. _if and only if_ Fix(h) _is almost tame_.

LEMMA 2 (Moise [8]). _If_ h _is a periodic homeomorphism of_ S^3 _with_ Fix(h) _a tame simple closed curve, then_ h _is a_ P.L. _homeomorphism_.

PROOF OF THEOREM 1. If n = 1, the theorem follows trivially from Lemma 1 and Walhausen's theorem [19]. For n > 1, we note that for each integer k with $0 < k < 2n$, $\text{Fix}(h) \subset \text{Fix}(h^k)$. By either [7] of [15], $\mathrm{Fix}(h^k)$ is a simple closed curve. Thus, $\mathrm{Fix}(h) = \mathrm{Fix}(h^k)$ and, in particular, $\mathrm{Fix}(h) = \mathrm{Fix}(h^n)$. Since h^n is an involution, it follows from Lemma 1 and [19] that $\mathrm{Fix}(h)$ is tame. Hence, by Lemma 2, h is P.L. and by [19] conjugate to a standard rotation.

3. FREE ACTIONS OF $Z/2n$ ON S^3.

Since a free action on S^3 acts simplicially on some triangulation of S^3, [12] or [16], all objects in this section shall be considered from the polyhedral point of view. The interior and closure of a point set M in a space will be denoted by int(M) and cl(M), respectively. The symbol V_p will denote a standard handlebody of genus p in S^3 (E^3).

LEMMA 3. _Suppose_ Z/n _acts freely on_ V_p _and_ $h \in Z/n$ _a generator_. _If_ h _is orientation preserving, then_ $V_p/\langle h\rangle$ _is homeomorphic to_ $V_{(p+n-1)/n}$ _and_ $p = kn+1$ _for some non-negative integer_ k. _In particular_, _h is a standard rotation of_ V_p, _up to conjugation_.

PROOF. Let $\pi : V_p \to \mathcal{V} = V_p/\langle h\rangle$ denote the natural projection, D a handle cutting disc for V_p and $d = \partial D$. Then $\pi(d)$ is a homotopically non-trivial closed curve on $\partial\mathcal{V}$ which is not necessarily simple. However, $\pi(d)$ can be shrunk on $\pi(D)$ to a point in $\mathcal{V}$. It follows from Dehn's Lemma and the loop theorem [9], [10] or [17] that there is a non-singular disc $E' \subset \mathcal{V}$ such that $e' = E' \cap \partial\mathcal{V} = \partial E'$ is homotopically non-trivial on $\partial\mathcal{V}$ and is a conservative ε-alteration of $\pi(d)$, [4].

Let $x \in V_p$ such that $\pi(x) \in E'$ and let $i : E' \to \mathcal{V}$ denote the inclusion. Since $i_* \Pi_1(E', \pi(x)) = \{0\} \subset \pi_* \Pi_1(V_p, x)$ it follows from [14, p. 198 and p. 195] that $\pi^{-1}(E')$ is a disjoint collection of n discs which permute under h and each gets mapped homeomorphically onto E' by π.

Let E_1 be a component of $\pi^{-1}(E')$ and $N(E_1)$ a sufficiently small regular neighborhood of E_1 in V_p such that $\cup h^i(N(E_1))$ is a disjoint collection of 3-cells. Let $W_1 = cl(V_p - \cup h^i(N(E_1)))$ and consider the components of W_1. We note that W_1 might have more than one component but that they must again be handlebodies (possibly of genus 0). If W_1 is not the union of disjoint 3-cells, then we repeat out argument to obtain $W_2 = cl(W_1 - \cup h^i(N(E_2)))$, where E_2 has the additional property that $h^i(E_1) \cap E_2 = \emptyset$. This choice of E_2 is made possible by choosing a handle cutting disc D such that $h^i(E_1) \cap D = \emptyset$ and then using an appropriate ϵ-alteration in the loop theorem on $\pi(\partial D)$.

A finite repetition of this argument must eventually yield disjoint discs $E_1, \ldots, E_m$ such that $\cup_{i,j} h^i(E_j)$ cuts V_p into n 3-cells which permute under h. Thus, $\mathcal{V}$ may be viewed as being obtained from one of these 3-cells by identifying appropriate pairs of disjoint discs in the boundary of that 3-cell. Since h is orientation preserving, it follows that $\partial\mathcal{V}$ is orientable. Therefore, $\mathcal{V}$ is a handlebody. Furthermore, $\chi(\partial\ ) = \chi(\partial V_p/\langle h \rangle) = \chi(\partial V_p)/n = 2(1-p)/n$ is even since $\partial\mathcal{V}$ is orientable. Hence $p = kn+1$ and $\mathcal{V}$ has genus $(p+n-1)/n$. This proves Lemma 3.

In [13] we proved the following two theorems.

THEOREM 2. *If* $Z/2n$ *acts freely on* S^3 *and if* $h \in Z/2n$ *is a generator, then there is polyhedral handlebody* $V \subset S^3$ *such that* $h(V) = S^3 - \text{int}V$.

THEOREM 3. *If* $Z/2n$ *acts freely on* S^3 *and* $h \in Z/2n$ *is a generator such that* h^2 *is topologically equivalent to an orthogonal transformation, then* h *is topologically equivalent to an orthogonal transformation*.

Our next result is now an easy consequence.

THEOREM 4. *Every free action of* $Z/2n$ *on* S^3 *is topologically equivalent to a standard orthogonal transformation. In particular, if* $Z/2n$ *acts freely on* S^3*, then* $S^3/(Z/2n)$ *is homeomorphic to a lens space of type* $L(2n)$.

PROOF. Let $h \in Z/2n$ be a generator. By Theorem 2 there is a handlebody $V \subset S^3$ such that $h(V) = S^3 - \text{int}V$. Thus, the pair $(V, h(V))$ is a Heegaard splitting of S^3. By Waldhausen's result [18] we may view V as V_p for some p. By Lemma 3, h^2 is a standard rotation on the pair $(V_p, h(V_p))$ and, hence, on S^3. Thus, by Theorem 3, h is topologically equivalent to an orthogonal transformation.

REMARK. Since submission of this paper the proof of a key Theorem of [13] for the $Z/2^k$ case had to be slightly modified, resulting in a direct classification of all free $Z/2n$ actions.

REFERENCES

1. R. H. Bing, A homeomorphism between the 3-sphere and the sum of two solid horned spheres, Ann. of Math. 56 (1952), 354-362.

2. R. H. Bing, Inequivalent families of periodic homeomorphisms of E^3, Ann. of Math. 80 (1964), 79-93.

3. S. Eilenberg, On the problems of topology, Ann. of Math. 50 (1949), 247-260.

4. D. W. Henderson, Extension of Dehn's lemma and the loop theorem, Trans. Amer. Math. Soc. 120 (1965), 448-469.

5. K. W. Kwun and J. L. Tollefson, Involutions of 3-manifolds with almost tame fixed point sets are PL, Houston J. of Math. (to appear).

6. G. R. Livesay, Fixed point free involutions of the 3-sphere, Ann. of Math. 72 (1960), 603-611.

7. E. E. Moise, Periodic homeomorphisms of the 3-sphere, Ill. J. of Math. 6 (1962), 206-225.

8. E. E. Moise, Periodic homeomorphisms of the 3-sphere with tame fixed point sets (to appear).

9. C. D. Papakyriakopoulos, On solid tori, Proc. London Math. Soc. 7 (1957), 281-299.

10. C. D. Papakyriakopoulos, On Dehn's lemma and the asphericity of knots, Ann. of Math. 66 (1957), 1-26.

11. P. M. Rice, Free actions of Z/4 on S^3, Duke Math. J. 36 (1969), 749-750.

12. G. X. Ritter, Free Z/8 actions on S^3, Trans. Amer. Math. Soc 181 (1973), 192-212.

13. G. X. Ritter, Free actions of Z/2n on S^3, (to appear).

14. H. Seifert and W. Threlfall, Lehrbuch der Topologie, Teubner Verlag, Leipzig, 1934.

15. P. A. Smith, Transformations of finite period, Ann. of Math. 39 (1938), 127-164.

16. P. A. Smith, Periodic transformations of 3-manifolds, Ill. J. of Math. 9 (1965), 343-348.

17. J. Stallings, On the loop theorem, Ann. of Math. 72 (1960), 12-19.

18. F. Waldhausen, Heegaard-Zerlegungen der 3-Sphäre, Topology 7 (1968), 195-203.

19. F. Waldhausen, Über Involutionen der 3-Sphäre, Topology 8 (1969), 81-92.

UNIVERSITY OF FLORIDA,
GAINSEVILLE, FL 32611,
U.S.A.

PERIODIC HOMEOMORPHISMS ON NON-COMPACT 3 MANIFOLDS

GERHARD X. RITTER AND BRADD E. CLARK

ABSTRACT

It has been conjectured that all periodic homeomorphisms acting piecewise linearly on a given 3-manifold must be conjugate to the standard periodic homeomorphisms on that manifold. In this paper we will consider the non-compact manifolds $R^1 \times T^2$, $R^2 \times S^1$, and $R^1 \times S^2$. We show that if h is a fixed point free involution, then $(R^1 \times T^2)/h$ is homeomorphic to either $R^1 \times T^2$ or $R^1 \times K^2$, that $(R^2 \times S^1)/h$ is homeomorphic to either $R^2 \times S^1$ or the open solid Klein bottle, and that $(R^1 \times S^2)/h$ is homeomorphic to either $R^1 \times P^2$ or the open 3-dimensional Möbius band. An easy consequence is that for any integer $n \geq 1$, $R^2 \times S^1$ admits only the obvious two free actions of period 2^n.

1. INTRODUCTION

It has been conjectured that all periodic homeomorphisms acting piecewise linearly on a given 3-manifold must be conjugate to the standard periodic homeomorphisms on that manifold. Much work has been done on this question when the given 3-manifold is compact. The question naturally arises as to whether or not analogous results may be obtained for various non-compact 3-manifolds

We will answer this question affirmatively for several such manifolds in this paper.

Since it follows from either [5] or [7] that free actions on a 3-manifold may be considered as acting piecewise linearly on some triangulation of that 3-manifold, we shall view all objects throughout this paper as being piecewise linear with respect to a given triangulation of that manifold.

We shall let R^k, S^k, T^k, P^k and K^k stand for k-dimensional Euclidean space, the k-dimensional sphere, torus, projective plane and Klein bottle, respectively. A homeomorphism of h of period n on the 3-manifold M is called a <u>free action</u> on M if for every positive integer $k < n$, h^k is a fixed point free homeomorphism of M onto itself. If M is a manifold, then the interior of M will be denoted by int M and the boundary of M by ∂M.

2. FIXED POINT FREE INVOLUTIONS ON $R^1 \times T^2$.

The goal of this section is the classification of the space $(R^1 \times T^2)/h$ when h is a free action of period two on $R^1 \times T^2$. In particular, we shall show that $(R^1 \times T^2)/h$ is homeomorphic to either $R^1 \times T^2$ or $R^1 \times K^2$. We will view $R^1 \times T^2$ as $(-1,1) \times T^2$.

LEMMA 2.1. <u>If</u> h <u>is a fixed point free involution on</u> $R^1 \times T^2$, <u>then there is a torus</u> T <u>in</u> $R^1 \times T^2$ <u>which is isotopic to</u> $\{0\} \times T^2$ <u>and such that either</u> $hT = T$ <u>or</u> $hT \cap T = \emptyset$.

PROOF. Let $T = \{0\} \times T^2$. If $hT \neq T$ and $T \cap hT \neq \emptyset$, then by using small isotopic deformations whenever necessary, we may assume that $T \cap hT$ consists of at most a finite number of simple closed curves.

If J is a simple closed curve in $T \cap hT$ such that J bounds a disk D on T with the property that $D \cap hT = J$, then we shall call J an <u>innermost</u> simple closed curve on T and D an <u>innermost</u> disk on T. Innermost simple closed curves and disks on hT are defined in an analogous fashion.

We shall say that a component J of $T \cap hT$ is of <u>type</u> (i), (ii), or (iii) if:

(i) J is homotopically trivial on both T and hT.

(ii) J is homotopically trivial on one of T or hT, but not both.

(iii) J is homotopically non-trivial on both T and hT.

Suppose J is of type (i) and innermost on hT. Let T denote the dist on hT bounded by J and let E denote the disk on T bounded by J. We choose a simple closed curve $J' \subset T-E$, sufficiently close to J, such that the annulus A on T with $\partial A = J \cup J'$ has the property that $A \cap hT = J$. Now choose a disk D′ so close to D that D′ satisfies $D' \cap T = \partial D' = J'$ and $D' \cap hT = D' \cap hD' = \emptyset$. This choice of D′ is possible since D is innermost and h is fixed point free.

Since $D' \cup D \cup A$ bounds a 3-ball in $R^1 \times T^2$, the torus $T' = [T-(E \cup A)] \cup D'$ is isotopic to T in $R^1 \times T^2$. It follows that $T' \cap hT'$ is a strict subset of $T \cap hT$ and contains fewer components of type (i). A finite number of repetitions of this process will yield a torus T″ which is isotopic to $\{0\} \times T^2$ and has the property that no component of $T'' \cap hT''$ is of type (i).

For the sake of convenience, we shall again denote the adjusted torus T'' by T. Now suppose that a component J of $T \cap hT$ is of type (ii) and bounds a disk D on hT. If D is not innermost on hT, then there must be a component J' of $T \cap hT$ with $J' \subset D$ and J' innermost on hT. Since J' cannot be of type (i), it is of type (ii). Thus, if D' denotes the disk on hT bounded by J', then $D' \cap T = J'$ with J' non-trivial on T. But this is impossible since T is incompressible in $R^1 \times T^2$.

If J is trivial on T and non-trivial on hT, then hJ is non-trivial on T and trivial on hT. But this is impossible by the above argument. Therefore, no component of $T \cap hT$ is of type (ii). Thus either $T \cap hT = \emptyset$ or all components of $T \cap hT$ are of type (iii).

We now suppose that all components of $T \cap hT$ are of type (iii). If n is the number of components of $T \cap hT$, then $T \cap hT$ divides hT into n annuli, $A_1, \ldots, A_n$, such that $T \cap \text{int } A_i = \emptyset$. Each annulus A_i satisfies one and only one of the following conditions:

(1) $\partial A_i \cap \partial hA_i = \partial A_i$

(2) $\partial A_i \cap \partial hA_i = \emptyset$

(3) $\partial A_i \cap \partial hA_i$ contains exactly one component of $T \cap hT$.

We consider each of these possibilities separately.

(1) Suppose $A_i \subset hT$ is an annulus satisfying (1). Then ∂A_i divides T into two annuli, B_1 and B_2. Since $T \cap \text{int } A_i = \emptyset$, we may assume without loss of generality that $hA_i = B_1$. Let $T_1 = A_i \cup B_1$ and $T_2 = A_i \cup B_2$. Then one of the tori T_1 or T_2 does not bound a solid torus in $R^1 \times T^2$ and must be isotopic to T.

Suppose that T_2 bounds a solid torus V in $R^1 \times T^2$. Let M be any polyhedron containing B_1 and such that $M \cap V = \partial B_2 = \partial B_1$. Then it follows from section 8 of [6] that there is an isotopy which takes A_i onto B_2 and is the identity on M. Hence T_1 is isotopic to T.

If T_1 is isotopic to T, then T satisfies the conclusion of Lemma 2.1 since $hT_1 = h(A_i \cup B_1) = B_1 \cup A_i = T_1$. If T_2 is isotopic to T, let J and K denote the components of ∂A_i. We may choose a simple closed curve J' and K' on int B_2, sufficiently close to J and K respectively, such that the annuli R_1 and R_2 on B_2 bounded by J, J' and K, K' respectively, have the property that $hT \cap R_1 = J$ and $hT \cap R_2 = K$. Let A_i' be an annulus sufficiently close to A_i so that A_i' satisfies:

(a) $A_i' \cap T = A_i' \cap B_2 = \partial A_i' = J' \cup K'$

(b) $A_i' \cap hT = A_i' \cap hA_i' = \emptyset$

(c) $T' = [B_2 - (R_1 \cup R_2)] \cup A_i'$ is isotopic to T.

We can obtain condition (b) since $T \cap \text{int } A_i = hA_i' \cap T = \emptyset$ and $hA_i' \cap hT = h\partial A_i'$. Also since $T_2 = A_i \cup B_2$ is isotopic to $\{0\} \times T^2$, we can fulfill condition (c) by taking A_i' close enough to A_i.

By our construction of T', the number of components of $T' \cap hT'$ is strictly less than the number of components of $T \cap hT$. Using this method we can obtain a torus T'' such that either $T'' = hT''$ or $T'' \cap hT''$ are all of type (iii) with no annulus $A_i \subset hT''$ satisfying (1) $\partial A_i \cap \partial hA_i = \partial A_i$.

(2) We suppose that no annulus $A_i \subset hT$ satisfies (1) and that for some i, A_i satisfies condition (2) $\partial A_i \cap \partial hA_i = \emptyset$. Then

∂A_i divides T into two annuli B_1 and B_2. Since $T \cap \text{int } A_i = \partial A_i \cap \partial hA_i = \emptyset$ we may suppose without loss of generality that $hA_i \subset \text{int } B_1$. Let $T_1 = A_i \cup B_1$, $T_2 = A_1 \cup B_2$, and J and K denote the boundary components of A_i. As before, at least one of the tori T_1 or T_2 is isotopic to T and hence to $\{0\} \times T^2$.

If T_1 is isotopic to $\{0\} \times T^2$, let J′ and K′ be two simple closed curves on int B_1, close to J and K respectively, and A_i' an annulus sufficiently close to A_i fulfilling the conditions:

(a) $A_i' \cap T = A_i' \cap B_1 = \partial A_i' = J' \cup K'$

(b) $A_i' \cap hT = A_i' \cap hA_i' = \emptyset$

(c) If R_1 and R_2 denote the annuli on B_1 bounded by J, J′ and K, K′ respectively, then $hT \cap R_1 = J$, $hT \cap R_2 = K$

(d) $T' = [B_1 - (R_1 \cup R_2)] \cup A_i'$ is isotopic to T.

As in argument 1, all of these conditions can be satisfied. By construction, the number of components of $T' \cap hT'$ is strictly less than the number of components of $T \cap hT$.

If T_2 is isotopic to $\{0\} \times T^2$, then J and K are not components of $T_2 \cap hT_2$. Either way, we can by repeated application of the above algorithm find a torus T such that T is isotopic to $\{0\} \times T^2$ and either $hT = T$ or $T \cap hT$ consists of a finite number of simple closed curves dividing hT into annuli all satisfying condition (3), ($\partial A_i \cap \partial hA_i$ contains exactly one component of $T \cap hT$).

(3) Let $A_i \subset hT$ be an annulus with $J = A_i \cap hA_i$. Also, let A_j denote the annulus on hT adjacent to A_i with $A_i \cap A_j = J$. Let J_i and J_j denote the remaining boundary components of A_i and A_j respectively. The curves J and J_i divide T into two annuli B_1 and

B_2. We may assume that $hA_i \subset B_1$, and therefore $hA_j \subset B_2$. The set $J \cup J_j$ also divides T into two annuli B_1' and B_2'. We may assume that $hA_i \subset B_1'$.

Again at least one of the tori $T_1 = A_j \cup B_1'$ or $T_2 = A_j \cup B_2'$ is isotopic to $\{0\} \times T^2$. If T_1 is isotopic to $\{0\} \times T^2$, let J_j' and J' be two simple closed curves on int B_1' close to J_j and J respectively. Let A_j' be an annulus with $\partial A_j' = J_j' \cup J'$, sufficiently close to A_j to fulfill the conditions:

(a) $A_j' \cap T = A_j' \cap B_1 = J_j' \cup J'$

(b) $A_j' \cap hT = A_j' \cap hA_j' = \emptyset$

(c) If R_1 and R_2 denote the two annuli on B_1' with boundary components J, J' and J_j, J_j' respectively, then $hT \cap R_1 = J$ and $hT \cap R_2 = J_j$

(d) $T' = [B_1' - (R_1 \cup R_2)] \cup A_j'$ is isotopic to $\{0\} \times T^2$.

On the other hand, if T_2 is isotopic to $\{0\} \times T^2$, then so is at least one of $B_2 \cup A_i$ and $(B_2 - \text{int } B_2') \cup A_i \cup A_j$. If $B_2 \cup A_i$ is isotopic to $\{0\} \times T^2$ we find simple closed curves J' and J_i' on int B_2, close to J and J_i respectively. We find an annulus A_i' sufficiently close to A_i to insure that

(a) $A_i' \cap T = A_i' \cap B_2 = \partial A_i' = J' \cup J_i'$

(b) $A_i' \cap hT = A_i' \cap hA_i' = \emptyset$

(c) If R_1 and R_2 denote the two annuli on B_2 bounded by J, J' and J_i, J_i' respectively, then $hT \cap R_1 = J$, and $hT \cap R_2 = J_i$

(d) $T' = [B_2 - (R_1 \cup R_2)] \cup A_i'$ is isotopic to $\{0\} \times T^2$.

Finally if $(B_2 - \text{int } B_2') \cup A_i \cup A_j$ is isotopic to $\{0\} \times T^2$, we simply set $T' = (B_2 - \text{int } B_2') \cup A_i \cup A_j$ and note that since $T' \cap h(J_i \cup J_j) = \emptyset$, $T' \cap hT'$ contains fewer components than $T \cap hT$. By repeated use of this algorithm we can find a torus T which is isotopic to $\{0\} \times T^2$ and such that either $hT = T$ or $T \cap hT = \emptyset$.

THEOREM 2.2. _If_ h _is a fixed point free involution on_ $R^1 \times T^2$, _then there is a torus_ T _in_ $R^1 \times T^2$ _which is isotopic to_ $\{0\} \times T^2$ _and such that_ $hT = T$.

PROOF. By Lemma 2.1 we may assume that we have found a torus T isotopic to $\{0\} \times T^2$ and with $T \cap hT = \emptyset$. Let Y denote the closure of the component of $R^1 \times T^2 - (T \cup hT)$ which is homeomorphic to $T^2 \times (0,1)$. We may write Y as $T^2 \times [0,1]$ and $T \cup hT = T^2 \times \{0,1\}$.

Theorem 10.3 of [2] states that if F is a compact connected 2-manifold other than S^2, a 2-cell, and the projective plane; with $\tau : F \times [0,1] \to F \times [0,1]$ a free involution and $\tau(F \times \{0,1\}) = F \times \{0,1\}$, then τ is equivalent relative to $F \times \{0,1\}$ to a free involution $\tau_1 : F \times [0,1] \to F \times [0,1]$ such that either

(i) $\tau_1(x,t) = (\sigma(x),t)$ or

(ii) $\tau_1(x,t) = (\sigma(x),1-t)$ for some free involution $\sigma : F \to F$.

We let $\tau = h$ and T be the image of $\{1/2\} \times T^2$ under the conjugation homeomorphism. Then $hT = T$.

THEOREM 2.3. _If_ h _is a fixed point free involution on_ $R^1 \times T^2$, _then the orbit space_ $(R^1 \times T^2)/h$ _is homeomorphic to either_ $R^1 \times T^2$ _or_ $R^1 \times K^2$.

PROOF. By Theorem 2.2 we have a torus T isotopic to

$\{0\} \times T^2$ and invariant under h. T divides $R^1 \times T^2$ into two components, V and W. We may think of V and W as $[0,1] \times T^2$. If $hV = W$, then $(R^1 \times T^2)/h$ may be viewed as being obtained from $[0,1) \times T^2$ by identifying $\{0\} \times T^2$ by h. It follows that $(R^1 \times T^2)/h$ is homeomorphic to either $R^1 \times T^2$ or $R^1 \times K^2$ since $(\{0\} \times T^2)/h$ is either T^2 or K^2.

Now suppose that $hV = V$. Since V-T is homeomorphic to $R^1 \times T^2$ and $h(V-T) = V-T$; by theorem 2.2 we can find a torus T_1 such that $T_1 \subset V-T$ and $hT_1 = T_1$. Also by E. M. Brown's work [1], we may choose the product structure of $[0,1) \times T^2$ so that $T_1 = \{1/2\} \times T^2$. Since $(1/2, 1) \times T^2$ is homeomorphic to $R^1 \times T^2$ we may continue this process until we have a sequence of tori $\{T_n\}$ such that

$$\text{(i)} \quad hT_n = T_n = \{n/(n+1)\} \times T^2$$

and a sequence of toroidal shells $\{V_n\}$ such that

$$\text{(ii)} \quad hV_n = V_n = [n/n+1, \ (n+1)/(n+2)] \times T^2 \text{ and } \cup V_n = V.$$

A similar argument gives us a sequence of toroidal shells $\{W_n\}$ with $hW_n = W_n$ and $\cup W_n = W$.

We again use Hempel's theorem [2] on each of these toroidal shells and obtain for the orbit space of $([n/(n+1), \ (n+1)/(n+2)] \times T^2)$ either $[n/(n+1), \ (n+1)/(n+2)] \times T^2$ or $[n/(n+1), \ (n+1)/(n+2)] \times K^2$. This is because $(\{n/(n+1)\} \times T^2)/h$ is either T^2 or K^2. Therefore $(R^1 \times T^2)/h$ is either $R^1 \times T^2$ or $R^1 \times K^2$.

3. FIXED POINT FREE INVOLUTIONS ON $R^2 \times S^1$.

In this section we shall consider the space $(R^2 \times S^1)/h$

where h is a free action of period 2 on $R^2 \times S^1$. We shall show that $(R^2 \times S^1)/h$ is homeomorphic to either $R^2 \times S^1$ or the open solid Klein bottle.

We begin with the following lemma.

LEMMA 3.1. _If_ h _is a fixed point free involution on_ $R^2 \times S^1$, _then there exists a solid torus_ $V \subset R^2 \times S^1$, _such that if_ C _is a core of_ V, C _is also a core of_ $R^2 \times S^1$; _and_ $hV = V$.

PROOF. Let C be a core of $R^2 \times S^1$. By using small isotopic deformations whenever necessary, we may assume that $hC \cap C = \emptyset$. Let A be a non-singular annulus such that $\partial A = C \cup hC$. Then $A \cup hA$ is a singular torus in $R^2 \times S^1$.

We note that $(C \cup hC) \cap (\text{int } A \cap \text{int } hA) = \emptyset$. Thus by the techniques of Lemma 2.1 we can find a torus T' such that $C \subset T'$ and either $hT' = T'$ or $hT' \cap T' = \emptyset$. If the second case is the result, by Theorem 2.2 we can still find a torus T with $hT = T$ and T is isotopic to T' in $R^2 \times S^1$. Let V be the solid torus bounded by T. Then $hV = V$ and any core of V is a core of $R^2 \times S^1$.

THEOREM 3.2. _If_ h _is a fixed point free involution on_ $R^2 \times S^1$ _then_ $(R^2 \times S^1)/h$ _is homeomorphic to either_ $R^2 \times S^1$ _or the open solid Klein bottle_.

PROOF. Let $V \subset R^2 \times S^1$ be the solid torus of Lemma 3.1. Since $hV = V$, we know that V/h is either a solid torus or a solid Klein bottle. Also, we note that $h((R^2 \times S^1)-V) = (R^2 \times S^1)-V$ and $(R^2 \times S^1)-V$ is homeomorphic to $R^1 \times T^2$. By Theorem 2.3 we know that $(R^2 \times S^1)-V/h$ is homeomorphic to either $R^1 \times T^2$ or $R^1 \times K^2$. Since these results must agree on ∂V, we have that $(R^2 \times S^1)/h$ is homeomorphic to either $R^2 \times S^1$ or the open solid Klein bottle.

If $Z/2^n$ acts freely on $R^2 \times S^1$ and $h \in Z/2^n$ a generator such that h^2 is a standard action, then there is a solid torus V such that $h^2V = V$ and V satisfies the conclusion of Lemma 3.1. If $T = \partial T$ and $hT \neq T$, then we can use our previous argument to obtain a torus T′ such that $h^2T' = T'$, T′ has the same essential properties as T and either $hT' = T'$ or $hT' \cap T' = \emptyset$. If $T' \cap hT' = \emptyset$, then either $T' \subset \text{int } hV'$ or $hT' \subset \text{int } V'$, where V′ denotes the solid torus bounded by T′. Applying h to the first relation, we obtain the amusing contradiction $hT' \subset \text{int } V'$. Similarly, $hT' \subset \text{int } V'$ cannot hold. Therefore $hT' = T'$. Thus, using straightforward induction, we obtain

COROLLARY 3.3. <u>If</u> $Z/2^n$ <u>acts freely on</u> $R^2 \times S^1$, <u>then</u> $(R^2 \times S^1)/(Z/2^n)$ <u>is homeomorphic to either</u> $R^2 \times S^1$ <u>or the open solid Klein bottle.</u>

4. FIXED POINT FREE INVOLUTIONS OF $R^1 \times S^2$.

Let B^3 be a 3-ball in R^3. We define the 3-dimensional Möbius band to be the space obtained from $R^3 - \text{int } B^3$ by identifying antipodal points on $\partial(R^3 - \text{int } B^3) = S^2$.

In this section we wish to show that $(R^1 \times S^2)/h$ is homeomorphic to either $R^1 \times P^2$ or the open 3-dimensional Möbius band. We shall view $R^1 \times S^2$ as $(-1,1) \times S^2$ in this section.

LEMMA 4.1. <u>If</u> h <u>is a fixed point free involution of</u> $R^1 \times S^2$ <u>then there exists a 2-sphere</u> S <u>in</u> $R^1 \times S^2$ <u>which is isotopic to</u> $\{0\} \times S^2$ <u>and such that</u> $hS = S$.

PROOF. Let $S = \{0\} \times S^2$. If $hS \neq S$ and $S \cap hS \neq \emptyset$, then by using small isotopic deformations whenever necessary, we may assume that $S \cap hS$ consists of a finite number of disjoint simple closed curve.

We shall call a component J of $S \cap hS$ innermost on S if the disk $D \subset S$ bounded by J has the property that $D \cap hS = J$. D will in this case be called an innermost disk on S. An innermost curve and disk on hS is defined analogously.

Let J be an innermost curve on hS and $D \subset hS$ be the innermost disk with $\partial D = J$. Also let D_1 and D_2 denote the two disks on S whose common boundary is J. We may assume that $hD \subsetneq D_1$. Since S does not bound a 3-ball in $R^1 \times S^2$, at least one of $S_1 = D \cup D_1$ or $S_2 = D \cup D_2$ cannot bound a 3-ball in $R^1 \times S^2$. We consider these cases separately.

Case 1. $S_1 = D \cup D_1$ does not bound a 3-cell. If $hJ = J$, then $hS_1 = h(D \cup D_1) = D_1 \cup D = S$. If $hJ \neq J$, we choose a simple closed curve $J' \subset D_1$ sufficiently close to J such that the annulus $A \subset D_1$ bounded by J and J' has the property that $A \cap hS = J$. Let D' be a disk sufficiently close to D such that $D' \cap hD' = D' \cap hs = \emptyset$, $D' \cap S = \partial D' = J$, and $S_1' = (D_1 - A) \cup D'$ does not bound a 3-ball in $R^1 \times S^2$. By construction, $S_1' \cap hS_1'$ is a strict subset of $S \cap hS$.

Continuing in this fashion we can find a two sphere S' such that $hS' \cap S' = \emptyset$ and S' does not bound a 3-cell. By the Combinatorial Annulus Theorem [8] we know that the region between S' and hS' is homeomorphic to $[-1/2, 1/2] \times S^2$. Also $h([-1/2, 1/2] \times S^2) = [-1/2, 1/2] \times S^2$ where $S' = \{-1/2\} \times S^2$ and

$hS' = \{1/2\} \times S^2$.

We may extend $h|[-1/2, 1/2] \times S^2$ to a fixed point free involution of S^3 in the natural fashion by attaching 3-cells to the boundary of $[-1/2, 1/2] \times S^2$. It now follows from [3] that there exists a 2-sphere S separating $\{-1/2\} \times S^2$ from $\{1/2\} \times S^2$ such that $hS = S$.

<u>Case 2</u>. $S_2 = D \cup D_2$ does not bound a 3-cell. If $hJ \neq J$, then $S_2 \cap hS_2 = D_2 \cap hD_2$ is a strict subset of $S \cap hS$. If $hJ = J$, choose a simple closed curve $J' \subset D_2$ such that the annulus $A \subset D_2$, bounded by J and J′ has the property that $A \cap hS = J$. Now choose a disk D′ sufficiently close to D such that

$D' \cap hD' = D' \cap hS = \emptyset$, $D' \cap S = \partial D' = J'$ and $S_2 = (D_2 - A) \cup D'$

does not bound a 3-ball in $R^1 \times S^2$.

By repeating this argument we can find a 2-sphere S′ which does not bound a 3-cell in $R^1 \times S^2$ and such that $S^1 \cap hS' = \emptyset$. Thus we can again use the argument of Case 1 to find a 2-sphere S such that $hS = S$ and S is isotopic to $\{0\} \times S^2$.

Henceforth, we let $S \subset R^1 \times S^2$ denote the 2-sphere of Lemma 4.1 and M_i, $i = 1,2$, the closed complementary domains of S in $R^1 \times S^2$. Since S is isotopic to $\{0\} \times S^2$ in $R^1 \times S^2$, M_i is homeomorphic to $[0,1) \times S^2$ for $i = 1,2$.

LEMMA 4.2. <u>If</u> $hM_i = M_i$, <u>then there is an open annulus</u> $A \subset R^1 \times S^2$ <u>such that</u> $hA = A$, $A \cap S$ <u>is a simple closed curve and</u> A <u>separate</u> $R^1 \times S^2$ <u>into two components, each homeomorphic to</u> $R^1 \times B^2$.

PROOF. Let R^3 be obtained from M_1 by filling in $\partial M_1 = S$

with a 3-ball. Then h extends naturally to an involution of R^3 with one fixed point. We may extend h further to $R^3 \cup \{\infty\} = S^3$ by setting $h(\infty) = \infty$. By [4], there is a 2-sphere $S^2 \subset S^3$ orthogonal to S, invariant under h and containing the two fixed points. We set $A_1 = S^2 \cap M_1$ and define $A_2 \subset M_2$ in a similar fashion. Also, since $h\partial A_1 = \partial A_1 \subset S = \partial M_2$, we may choose A_2 such that $\partial A_1 = \partial A_2$. The open annulus $A = A_1 \cup A_2$ satisfies the conclusion of Lemma 4.2.

THEOREM 4.3. _If_ h _is a fixed point free involution of_ $R^1 \times S^2$, _then_ $(R^1 \times S^2)/h$ _is homeomorphic to either_ $R^1 \times P^2$ _or the open 3-dimensional Möbius band_.

PROOF. If $hM_1 = M_2$, then $(R^1 \times S^2)/h$ may be viewed as being obtained from $[0,1) \times S^2$ by identifying antipodal points on the 2-sphere $\partial([0,1) \times S^2) = \{0\} \times S^2$. Thus $(R^1 \times S^2)/h$ is homeomorphic to the open 3-dimensional Möbius band.

If $hM_1 = M_1$, then by Lemma 4.2, there is an open annulus A which separates $R^1 \times S^2$ into two homeomorphic components N_1 and N_2. Since $hA = A$, either $hN_1 = N_1$ or $hN_1 = N_2$. However, since N is homeomorphic to $R^1 \times B^2$ and h is fixed point free of period 2, $hN_1 = N_1$ is not possible. Therefore $hN_1 = N_2$ and $(R^1 \times S^2)/h$ may be viewed as being obtained from $R^1 \times B^2$ by identifying antipodal points on $\partial(R^1 \times B^2) = R^1 \times \partial B^2$. Thus $(R^1 \times S^2)/h$ is homeomorphic to $R^1 \times P^2$.

REFERENCES

1. E. M. Brown, Unknotting in $M^2 \times I$, Trans. Amer. Math. Soc. 123 (1966), 480-505.
2. J. Hempel, 3-Manifolds, Ann. of Math. Studies, Princeton Univ. Press, Priceton, 1976.
3. G. R. Livesay, Fixed point free involutions on the 3-sphere, Ann. of Math. 72 (19609, 603-611.
4. G. R. Livesay, Involutions with two fixed points on the 3-sphere, Ann. of Math. 18 (1963), 582-593.
5. G. X. Ritter, Free Z/8 actions on S^3, Trans. Amer. Math. Soc. 181 (1973), 195-212.
6. H. Schubert, Knoten and Vollringe, Acta Math. 90 (1953), 131-286.
7. P. A. Smith, Periodic transformations of 3-manifolds, Illinois J. Math. 9 (1965), 343-348.
8. E. C. Zeeman, Seminar on combinatorial topology, Mimeographed Notes, Inst. Hautes Etudes Sci. Paris, 1963.

UNIVERSITY OF FLORIDA,

GAINSEVILLE, FL 32611,

U.S.A.

EQUIVARIANT FUNCTION SPACES AND EQUIVARIANT STABLE HOMOTOPY THEORY

REINHARD SCHULTZ

During the past six years relationships between spaces of equivariant self-maps of spheres and ordinary stable homotopy theory have been obtained by several different authors using somewhat different techniques [1, 2, 4, 6, 7, 9, 16]. Closer examination shows that such results fall into two classes: (i) Results relating equivariant stable homotopy theory to ordinary stable homotopy as in work of G. Segal, C. Kosniowski, T. tom Dieck, and H. Hauschild [4, 6, 7, 9, 16]. (ii) Results relating spaces of unpointed equivariant self-equivalences to homotopy theory as in the work of J. C. Becker and the author [1, 2].

This paper has two objectives. The first is to provide a natural relationship between the above classes of results, and the second is to apply this relationship to a question left open in [1, 2] - describing the composition product on the spaces F_G studied in those papers and describing (in principle, at least) how the Pontrjagin ring structure on $H_*(F_G)$ may be calculated. To be precise, we shall not use the homotopy equivalences constructed in [1, 2], but instead we shall replace them with more convenient maps defined in the same spirit.

It turns out that the systems studied in (i) and (ii) more

or less fit together as halves of a larger object; this is developed in Section 1. To describe this more fully, note that composition makes both types of systems into semigroups, and the resulting semigroups of arc components behave as follows: The systems of class (i) correspond to the multiplicative semigroup $|G|Z$ (assuming G is finite), while the systems of class (ii) correspond to the multiplicative semigroup $|G|Z + 1$. Of course, the disjoint union of these semigroups is again a semigroup, and it admits a considerable amount of extra algebraic structure. The vital features of this extra structure also exist on the spaces of interest to us (Section 2), and given this the objectives of the paper are reached by a straightforward topological translation of elementary algebraic identities as in [12, 13, 17].

ACKNOWLEDGEMENTS. This paper's title is very similar to those of two joint papers with James V. Becker [1, 2]. I am grateful to him for several discussions of the problems in the present paper. Comments and correspondence by A. Tsuchiya and H. Hauschild also proved to be constructive in setting up the formal framework of this paper. During the preparation of this paper I was partially supported by National Science Foundation Grant MPS 74-03609 and also MCS 76-08794.

1. THE UNREDUCED SUSPENSION

Throughout this paper G denotes a compact Lie group and V denotes a G-module (usually free, but not necessarily so at times). Following [1, 2] and [16], let S(V) and S^V denote the unit sphere in V and the one-point compactification of $V(= S(V \oplus R))$ respectively. As in [1], $F_G(V)$ denotes the space of G-equivariant self-maps of $S(V)$, and we define $F_G(S^V,\infty)$ to be the space of basepoint-preserving G-equivariant self-maps of S^V. The stable versions of these objects will be denoted by F_G and $F_G(\infty)$ respectively.

GENERAL REMARK. Completely analogous results hold if S^V is replaced by the unreduced suspension of a finite complex $K \simeq S^{\dim V-1}$ equipped with a free simplicial G-action (G finite). These generalizations parallel the extensions of [1] appearing in [2]; In fact, the techniques needed to write out the details of such generalizations are contained in [2].

The unreduced suspension functor induces continuous homomorphisms

(1.1)
$$\Sigma_V : F_G(V) \to F_G(S^V,\infty),$$

and it is an elementary exercise to check that the following diagram is homotopy-commutative:

(1.2)
$$\begin{array}{ccc} F_G(V) & \xrightarrow{\Sigma_V} & F_G(V,\infty) \\ {\scriptstyle \text{join with } S(W)}\downarrow & & \downarrow{\scriptstyle \text{smash with } S^W} \\ F_G(V\oplus W) & \xrightarrow{\Sigma_{V\oplus W}} & F_G(V\oplus W,\infty) \end{array}$$

(In fact, this homotopy can be realized by a specific ambient isotopy of $F_G(V \oplus W, \infty)$). It follows that there also is a stable analog of (1.1), $\Sigma : F_G \to F_G(\infty)$. The obvious conjectures are that Σ_V is roughly dim V-connected (on identity components at least) and Σ is a homotopy equivalence.

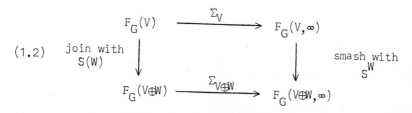

We shall prove these assertions using the spectral sequences of [14] *assuming* V *is a free module*. Recall from [14, §2] the spectral sequence

$$(1.3) \qquad E^2_{i,j} = H^{-i}(M(V),\ \pi_j(S(V))) \Rightarrow \pi_{i,j}(F_G(V)),$$

where $M(V) = S(V)/G$ as in [1]. The methods of [14] also yield a spectral sequence

$$(1.4) \qquad E^2_{p,q} = H^{-p}(\Sigma M(V),\ S^0;\ \pi_q(S^V)) \Rightarrow \pi_{p+q}(F^1_G(V,\ \infty)),$$

where Σ denotes unreduced suspension and F^1_G denotes the subspace of self-maps that are the identity on the fixed point set S^0. To obtain (1.4), replace F^1_G by the homotopically equivalent submonoid of all self-maps collapsing invariant disk neighborhoods of the poles into the latter points. The latter space equals the space of sections of the bundle $S(V) \times I \times_G S^V \to M(V) \times I$ that are constant on $M(V) \times \partial I$, and accordingly a relative version of [14, §§1 - 2] yields (1.4). In fact, the same sort of argument used to prove [14, Thm. 3.2] also gives the following basic comparison results:

PROPOSITION 1.5. *In the above notation,* Σ_V *induces a map from spectral sequence* (1.3) *to* (1.4). *On the* E^2 *level it is the map making the following diagram commute*:

$$\begin{array}{ccc} H^i(M(V);\ \pi_j(S(V))) & \xrightarrow{\ \Sigma\ } & H^{i+1}(\Sigma M(V),\ S^0;\ \pi_{j+1}(S^V)) \\ \downarrow{\scriptstyle\sigma} & \nearrow{\scriptstyle E_*} & \\ H^{i+1}(\Sigma M(V), S^0;\ \pi_j(S(V))) & & \end{array}$$

Notation: σ is the cohomological suspension isomorphism, while E_* is the coefficient map induced by the reduced suspension homomorphism $E: \pi_j(S(V)) \to \pi_{j+1}(S^V)$.

Since E is an isomorphism for $j \leq 2 \dim V - 2$ and an

epimorphism for $j = 2 \dim V-1$, the comparison theorem for spectral sequences (e.g., [18, Thm. 1]) and some elementary arithmetic yield the following conclusion:

COROLLARY 1.6. The map Σ_V is $(\dim V + \dim G)$-connected.

For the stable spaces F_G and $F_G(\infty)$ there is an obvious analog of Corollary 1.6:

COROLLARY 1.7. The stable unreduced suspension $\Sigma: F_G \to F_G(\infty)$ is a homotopy equivalence.

2. ALGEBRAIC OPERATIONS IN $F_G(S^V,\infty)$

Throughout this section V denotes a free G-module.

Consider the homomorphism $F_G(S^V,\infty) \to \{$basepoint-preserving self maps of $S^0\}$ given by restriction to the fixed point set. The codomain is a discrete space with two elements - the identity (written 1) and the constant map (written 0). The inverse images of these points determine two complementary clopen subsemigroups we shall call $F_G^1(S^V,\infty)$ and $F_G^0(S^V,\infty)$ respectively. We have already considered $F_G^1(S^V,\infty)$ in Section 1; on the other hand, the space $F_G^0(S^V,\infty)$ is an unstable approximation to the spaces arising in equivariant stable homotopy. In this section we shall describe the basic algebraic operations in $F_G(S^V,\infty)$ that bind these subspaces together. Of course, there is a similar splitting of $F_G(\infty)$ as $F_G^1(\infty) \cup F_G^0(\infty)$, and the results of this section have obvious (but generally unstated) stable analogs.

First of all, we recall the basic operations in the nonequivariant version of $F_G(S^V,\infty)$, which is merely $\Omega^{\dim V}(S^V)$.

The description as a monoid of self-maps yields a multiplication, the description as a loop space yields an addition, and stably these operations are homotopy distributive. Also, composition is stably homotopic to the external pairing induced by smash product. This algebraic structure is reflected in the identification of $\pi_0(\Omega^{\dim V}(S^V))$ with the ring of integers.

In our situation, multiplication is again given by composition, and the following diagram commutes up to a canonical homotopy:

$$\begin{array}{ccc} F_G(S^V,\infty)\times F_G(S^V,\infty) & \xrightarrow{\text{comp.}} & F_G(S^V,\infty) \\ {\scriptstyle\text{external smash}}\downarrow & & \downarrow{\scriptstyle \wedge 1_V} \\ F_G(S^{V\oplus V},\infty) & \xrightarrow{=} & F_G(S^{V\oplus V},\infty) \end{array}$$

(*Notation*: $\wedge 1_V$ means the smash product with the identity on S^V). The resulting monoid structure on the set of arc components is given by an elementary (but generally little-used) object:

PROPOSITION 2.1. *If* G *is finite, then* $\pi_0(F_G(S^V,\infty))$ *is isomorphic as monoids to the monoid of all integers congruent to* 0 *or* 1 $\operatorname{mod}|G|$. *If* G *is infinite, then* $\pi_0(F_G(S^V,\infty))$ *is isomorphic to the monoid* $\{0, 1\}$. *In both cases the arc components are specified by the degree of a representative*. *Furthermore, the components of* $F_G^{\epsilon}(S^V,\infty)$ *for* $\epsilon = 0, 1$ *corresponding to the set of all* $k \equiv \epsilon \bmod G$ (*finite case*) *or* $k = \epsilon$ (*infinite case*).

PROOF. (Sketch) Start out with an arbitrary $f \in F_G^{\epsilon}(S^V,\infty)$, where $\epsilon = 1$ or 0, and let e be the identity if $\epsilon = 1$ or the constant map if $\epsilon = 0$. Consider the obstructions to finding an equivariant homotopy from f to e that is fixed on S^0

(the obstruction theory connected with [14] or [15] is adequate for this purpose). The only nonzero obstruction $\mathcal{O}_G(f,e)$ lies in

$$H^{\dim S(V)+1}(\Sigma M(V),\ S^0;\ z) = z \quad \text{(finite) or } 0 \text{ (infinite)}$$

(twisted coefficients if $G = Z/2$ and dim V is odd), and functoriality properties of this obstruction imply

$$\deg e - \deg f = |G|\ \mathcal{O}_G(f,\ e).$$

Hence the arc component of f is completely determined by its degree, which is congruent (or equal) to 1 or 0. On the other hand, given e and an integer d, it is wellknown that an f can be found with $\mathcal{O}_G(f,e) = d$ (G finite!).

The key point of Proposition 2.1 is that addition in $\pi_0(F_G(S^V,\infty))$ is not globally definable. However, one can add elements of $\pi_0(F_G^0(S^V,\infty))$ to arbitrary elements of $\pi_0(F_G(S^V,\infty))$, and in favourable cases even subtraction is possible though addition may not be. This fragmentary additive structure may be realized topologically as follows:

DEFINITION. (a) The loop sum

$$*: F_G^0(S^V,\infty) \times F_G(S^V,\infty) \to F_G(S^V,\infty)$$

is given as follows: Consider the G-map

$$\psi_+ : S^V \to S^V \vee (S^V/S^0)$$

collapsing $S(V) \cup \{\infty\} \subseteq S^V$ to a single point, with ψ_+ projected onto S^V and S^V/S^0 being G-homotopic to the identity and the canonical quotient maps respectively. Explicitly, if $|v| \leq 1$ set $\psi_+(v) = (1-|v|^{-1})v \in S^V$, while if $|v| \geq 1$ set $\psi_+(v)$ = image of

$(\log |v|)\, v$ in S^V/S^0. Given $f \to F_G^0(S^V,\infty)$, let $\bar{f}: S^V/S^0 \to S^V$ be the map determined on the quotient space, and define $*(f, g) = f * g$ to be $(g + \bar{f})\psi_+$, where + means the combined map on the wedge.

(b) The <u>loop difference</u>

$$\diagdown\!\!\diagdown : \bigcup_{\epsilon=0,1} F_G^{\epsilon}(S^V,\infty) \times F_G^{\epsilon}(S^V,\infty) \to F_G^0(S^V,\infty)$$

is defined as follows: Let $S^V \vee_0 S^V$ denote the disjoint union of two copies of S^V mod identifying the zero points, and let the point at infinity in the first factor be the basepoint. Take $\psi_-: S^V \to S^V \vee_0 S^V$ to be the map defined by $\psi_-(v) = (\log |v|)\, v \in$ first factor if $|v| \geq 1$ and $\psi_-(v) = (1 - |v|)|v|^{-1} v \in$ the second factor if $|v| \leq 1$ (with self-evident conventions if $|v| = 0,\infty$). Given f and g, define $f \diagdown\!\!\diagdown g = (f + g)\psi_-$ as in (a).

(c) The <u>inverse map</u>

$$\chi: F_G^0(S^V,\infty) \to F_G^0(S^V,\infty)$$

is given by $\chi f = f\rho$, where $\rho(v) = |v|^{-2} v$.

It is elementary to check that the above operations have the following properties:

(2.2) <u>The maps</u> $(f, g, h) \to (f * g) * h$, $f * (g * h)$ <u>are homotopic</u>.

(2.3) <u>The maps</u> $f \to 0 * f$ <u>and</u> $g \to g * 0$ <u>are homotopic to the identity</u>.

(2.4) <u>The map</u> $f \to f \diagdown\!\!\diagdown f$ <u>is null-homotopic</u>.

(2.5) <u>If</u> $f \in F_G^0(S^V,\infty)$, <u>then the map</u> $f \to 0 \diagdown\!\!\diagdown f$ <u>is homotopic to</u> $\chi(f)$.

(2.6) χ^2 <u>is the identity</u>.

(2.7) <u>The maps</u> $(f, g, h) \to f * (g \diagdown\!\!\diagdown h)$, $(f * g) \diagdown\!\!\diagdown h$ <u>are homotopic</u>.

A homotopy-commutativity relation of the form $f * g \simeq g * f$

is notably absent from this list; we shall do this later (see Proposition 2.12).

The smash product and loop operations satisfy several standard homotopy-distributivity laws as in [12, 13, 18]:

(2.8) *The maps* $(f, g, h) \to f \wedge (g * h)$, $(f \wedge g) * (f \wedge h)$ *are homotopic; a similar statement holds if the order of the factors is reversed.*

(2.9) *The maps* $(f, g, h) \to f \wedge (g \setminus h)$, $(f \wedge g) \setminus (f \wedge h)$ *are homotopic, and likewise if the order of the factors is reversed.*

(2.10) *The maps* $(f, g) \to f \wedge (0 \setminus g)$, $\chi(f \wedge g)$, $\chi(f) \wedge g$, $f \wedge \chi(g)$ *are all homotopic, and likewise if the order of the factors is reversed.*

(2.11) *The maps* $f \to f \wedge 0$, $0 \wedge f$ *are nullhomotopic.*

Using the above elementary properties, we shall show that $*$ is stably homotopy-commutative. This result corresponds to the redundancy of the commutative law of addition in a ring with unit (e.g., [3, Exercise 1, p. 128]).

PROPOSITION 2.12. *Let* V *be as before, and let* W *be another free* G-*module of positive dimension; let* 1_W *be the identity on* S^W. *Then the maps* $(f, g) \to 1_W \wedge (f * g)$, $1_W \wedge (g * f)$ *are homotopic.*

PROOF. (Sketch). Consider the map

$$\Lambda: (f, g) \to (1_W \setminus 1_W) \wedge (f * g).$$

According to (2.2) - (2.11), Λ may be expanded two ways using homotopy distributivity. In particular, it follows that the maps

$$\lambda_1: (f, g) \to 1_W \wedge (f * (f * \chi(f)) * \chi(g))$$

$$\lambda_2: (f, g) \to 1_W \wedge (f * (\chi(f) * g) * \chi(g))$$

are homotopic. Consequently, the maps

$$\mu_1:(f, g) \to 1_W \wedge (f * \lambda_1(\chi(f), g) * g)$$

$$\mu_2:(f, g) \to 1_W \wedge (f * \lambda_2(\chi(f), g) * g)$$

are also homotopic. But it is immediate that $\mu_1(f, g) \simeq 1_W \wedge (g * f)$ and $\mu_2(f, g) \simeq 1_W \wedge (f * g)$.

More important for our purposes is the following close relationship between F_G^0 and F_G^1:

PROPOSITION 2.13. The map $(\ ) * 1_V: F_G^0(S^V, \infty) \to F_G^1(S^V, \infty)$ is a homotopy equivalence, and the map $(\ ) \diagdown 1_V$ is an explicit homotopy inverse.

3. COMPARISON OF ALGEBRAIC STRUCTURES

In the terminology of [13, p. 242], the operations of Section 2 make $F_G^0(\infty)$ into a Hopf ring without identity. Since $F_G^0(\infty)$ is homotopy equivalent to the spaces $Q(BG^{ad})$ where BG^{ad} denotes the Thom space of the vector bundle $EG \times_G \mathfrak{g} \to BG$ (see [1, 2] where ζ is used in place of ad), it is highly desirable to identify the induced Hopf ring structure on $Q(BG^{ad})$. If one studies this problem from the viewpoint of equivariant stable homotopy theory, it follows that the loop sum in $F_G^0(\infty)$ corresponds to the loop sum in $Q(BG^{ad}) = \Omega^\infty S^\infty(BG^{ad})$ and the smash product in $F_G^0(\infty)$ corresponds to the composite

$$Q(BG^{ad}) \times Q(BG^{ad}) \xrightarrow{\wedge} Q(BG^{ad} \wedge BG^{ad}) \xrightarrow{d_!} Q(BG^{ad}),$$

where $d_!$ is the umkehr map associated to the fiber bundle

$G \times G/G \subseteq BG \to BG \times BG$. In particular, these may be seen using the equivariant transversality - theoretic interpretation of $[X, F_G^O(\infty)]$ developed in [6, 7, 16]. However, for the sake of uniformity we shall outline a different proof using the methods of [1, 2]; in particular, this eliminates the need for a careful identification between the approaches of [1, 2] and [4, 6, 7, 9, 16]. Since the arguments are generally similar to those of [1], numerous details have been omitted and their verifications left to the reader.

We begin by defining a map $\lambda_O: F_G^O(S^V,\infty) \to Q(M(V)^{ad})$ analogous to the maps λ in [1, 2]. As in these papers, elements of $F_G^O(S^V,\infty)$ are equivalent to cross sections of the bundle $I \times S(V) \times_G S^V \to I \times M(V)$ whose restrictions to $\partial I \times M(V)$ equal the section $I \times S(V) \times_G \{\infty\}$; the latter in turn are equivalent to ex-maps $I \times M(V) \times S^O \to I \times S(V) \times_G S^V$ that are fixed on $\partial I \times M(V) \times S^O$. If we take the fiberwise suspension with the sphere bundle of $\nu_{M(V)}$-ad, we obtain maps $I \times S(\nu_M\text{-ad}) \to I \times M(V) \times S^n \to S^n$ that are constant on $I \times \{\text{cross section}\} \cup \partial I \times S(\nu_M\text{-ad})$; but the latter correspond to S-maps $M^{\nu\text{-ad}} \to S^O$, and S-dualization takes these into S-maps $S^O \to M(V)^{ad}$. All of these operations are easily checked to be continuous as in [1, 2], and therefore the construction described above yields a continuous map λ_O. The methods of [1, 2] imply that λ_O is approximately $(\dim V + \dim G\text{-}2)$-connected.

REMARKS 1. We shall <u>not</u> be using the maps λ constructed in [1, 2], and we do not assume any relationship between λ and λ_O except similarity of their definitions.

2. As in [1, 2], under the mappings λ_O the forgetful map $F_G^O(S^V,\infty) \to F_H^O(S^V,\infty)$ corresponds to the umkehr map $Q(M_G(V)^{ad}) \to Q(M_H(V)^{ad})$

where $M_K(V) = S(V)/K$ (compare Proposition 3.3 below).

The following observation is an immediate consequence of the definitions:

(3.1) <u>The map</u> λ_0 <u>preserves loop sums</u>.

Identification of the smash product is more difficult. If V and W are G-modules, the the direct sum $V \oplus W$ is a $G \times G$-module via the product of the two actions, and the usual G-module structure on $V \oplus W$ is the restriction of the action to the diagonal subgroup. In particular, the smash product map $F_G^0(S^V,\infty) \times F_G^0(S^W,\infty) \to F_G^0(S^{V\oplus W},\infty)$ may be factored through the space $F_{G\times G}^{00}(S^{V\oplus W},\infty)$ of all $G \times G$-equivarian maps that are trivial on $S^V \cup S^W \subseteq S^{V\oplus W}$. The following two results describe these factors in terms of λ_0.

PROPOSITION 3.2. There is a map $\lambda_{00}: F_{G\times G}^{00}(S^{V\oplus W},\infty) \to Q(M(V)^{ad} \wedge M(W)^{ad})$ <u>that is approximately</u> (dim V + dim W + 2 dim G-2)-<u>connected and makes the following square commute</u>:

$$\begin{array}{ccc} F_G^0(S^V,\infty) \times F_G^0(S^W,\infty) & \xrightarrow{\lambda_0 \times \lambda_0} & Q(M(V)^{ad}) \times Q(M(W)^{ad}) \\ \downarrow \text{smash} & & \downarrow \text{smash} \\ F_{G\times G}^{00}(S^{V\oplus W},\infty) & \xrightarrow{\lambda_{00}} & Q(M(V)^{ad} \wedge M(W)^{ad}) \end{array}$$

PROOF. Since the maps in F_G^{00} are trivial on $S^V \cup S^W$ and $G \times G$ acts freely on $S^{V\oplus W} - (S^V \cup S^W)$, it follows as above that elements of $F_G^{00}(S^{V\oplus W},\infty)$ correspond to sections of the bundle

$$D^2 \times S(V) \times S(W) \times_{G\times G} S^{V\oplus W} \to D^2 \times M(V) \times M(W)$$

that are constant on $S^1 \times M(V) \times M(W)$. Furthermore, this bundle is (a) the stable sphere bundle of $\tau_{M(V)\times M(W)} \oplus ad_{G\times G}$, and also

(b) the fiberwise smash product of $\pi^*_{M(V)}(S(V)\times_G S^V)$ with $\pi^*_{M(W)}(S(W)\times_G S^W)$. If we combine description (a) with the section-theoretic interpretation of $F^{OO}_G(S^{V\oplus W}, \infty)$, we may define the map λ_{OO} by exactly the same formal process (via ex-maps, fiberwise suspension to S-maps into S^0, S-dualization) used for λ_0. On the other hand, description (b) gives a strong hold on the section induced by a smash product of two maps - it is merely the smash product of the two sections. If we take the fiberwise suspension of such a smash product, a check of the definitions shows that the resulting S-map $M(V)^{\nu-ad} \wedge M(W)^{\nu-ad} \to S^0$ is merely homotopic to the smash product of the maps determined by the factors. Finally, S-dualization preserves smash products (up to homotopy, at least), and hence the above square is homotopy-commutative. The connectivity assertion follows from [1, Thm. 2.2] by the same considerations used in [1, p. 9].

PROPOSITION 3.3. _The following square is homotopy-commutative_:

$$\begin{array}{ccc}
F^{OO}_{G\times G}(S^{V\oplus W},\infty) & \xrightarrow{\ \lambda\ } & Q(M(V)^{ad} \wedge M(W)^{ad}) \\
 & & \big\downarrow d^{!} \\
\text{restriction to diagonal subgp.}\ \big\downarrow & & Q(S(V)\times_G S(W)^{ad}) \\
 & & \big\downarrow Q(\subseteq) \\
F^{O}_{G}(S^{V\oplus W},\infty) & \xrightarrow{\ \lambda\ } & Q(M(V\oplus W)^{ad})
\end{array}$$

The map $d^!$ is an umkehr map associated to the bundle $S(V)\times_G S(W) \to M(V)\times M(W)$, and $\subseteq$ refers to the injection onto the submanifold of all points with join coordinate 1/2 in $M(V\oplus W) \cong S(V) * S(W)/G$.

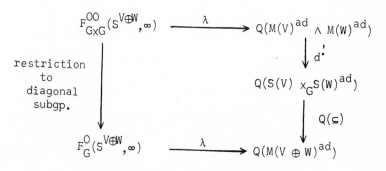

PROOF. The forgetful map on the left side of the above square may be described as follows: Given an equivariant map $D^2 \times S(V) \times S(W) \to S^{V \oplus W}$ that is constant on the boundary, a map $I \times S(V \oplus W) \to S^{V \oplus W}$ is obtained by viewing $I \times S(V \oplus W)$ as a quotient space of $D^2 \times S(V) \times S(W)$, (e.g., use the identification $S(V \oplus W) \cong S(V) * S(W) = S(V) \times S(W) \times I/$<u>eq. rel</u>.); a well-defined map is obtained because the original map is constant at all points where identifications are made. Thus we may factor the forgetful map in two steps - (a) pass from $G \times G$-equivariant maps $D^2 \times S(V) \times S(W)$ to G-equivariant maps, (b) pass from maps on $D^2 \times S(V) \times S(W)$ to maps on the quotient $I \times S(V \oplus W)$. Let $\mathcal{F}$ denote the intermediate space of G-equivariant maps $D^2 \times S(V) \times S(W) \to S^{V \oplus W}$ that are constant on the boundary.

By the same sort of construction used previously, there is a map $\bar{\lambda}: \mathcal{F} \to Q(S(V) \times_G S(W)^{ad})$ making the following square commute up to homotopy:

$$\begin{array}{ccc}
F_G^{00}(S^{V \oplus W}, \infty) & \xrightarrow{\lambda_{00}} & Q(M(V)^{ad} \wedge M(W)^{ad}) \\
\downarrow & & \downarrow d^{!} \\
\mathcal{F} & \xrightarrow{\bar{\lambda}} & Q(S(V) \times_G S(W)^{ad})
\end{array}$$

To prove commutativity, one needs a method for identifying the forgetful map with the umkehr map via λ_{00} and $\bar{\lambda}$. This may be done by a straightforward extension of the techniques used in [1, §§7-8].

Thus it is only necessary to show the following square is homotopy-commutative:

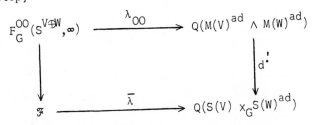

$$\begin{array}{ccc} \mathcal{F} & \xrightarrow{\bar{\lambda}} & Q((S(V) \times_G S(W))^{ad}) \\ \Big\downarrow \pi & & \Big\downarrow Q(\subseteq) \\ F_G(W^{V\oplus W}, \infty) & \xrightarrow{\lambda_0} & Q(M(V \oplus W)^{ad}) \end{array}$$

Recall that λ and λ_0 are obtained by constructing S-maps into S^0 from $(S(V) \times_G S(W))^{\nu\text{-}ad}$ and $M(V \oplus W)^{\nu\text{-}ad}$ respectively, and taking their S-duals. However, a check of the definitions shows that the S-map $M(V \oplus W)^{\nu\text{-}ad} \to S^0$ associated to some $\pi(f) \in F_G^0(S^{V\oplus W}, \infty)$ is obtainable by taking the S-map $(S(V) \times_G S(W))^{\nu\text{-}ad} \to S^0$ associated to $f \in \mathcal{F}$ and composing it with the collapsing map of Thom spaces given by the embedding $S(V) \times_G S(W) \subseteq M(V \oplus W)$ as the set of points with joint coordinate 1/2 (i.e., all $[v, w]$ with $|v| = |w|$, where $v \in V$ and $w \in W$). Since this collapsing map is S-dual to the Thom space map induced by inclusion, homotopy-commutativity of the square follows.

To conclude our treatment of the Hopf ring structure on $F_G^0(\infty)$, something should be said about stabilization. The techniques of [1] all go through to show that $\lambda_0 : F_G(S^V, \infty) \to Q(M(C)^{ad})$ has the same sort of stability properties proved in [1, (5.13), §§7-8]. Furthermore, these methods also show that all the maps and diagrams constructed in this section are functorial for inclusions $V \subseteq V'$, $W \subseteq W'$; verification of this is left to the reader.

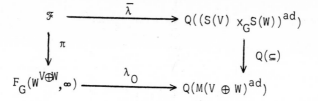

It follows that the maps λ_0 considered above have good stable analogs in the sense of [1, 2]; we shall abuse notation and also denote the stabilized maps $F_G^0(\infty) \to Q(BG^{ad})$ by λ_0. If it is ever necessary or desirable to make strong distinctions, symbols such as λ_{0V} will specify the unstable maps, and λ_0^{st} will specify the stable map.

The machinery developed above essentially reduces the description of composition on $F_G^1(\infty)$ and F_G to an easy exercise in arithmetic.

THEOREM 3.4. *Let* $\otimes: Q(BG^{ad}) \times Q(BG^{ad}) \to Q(BG^{ad})$ *be the map*

$$Q(BG^{ad}) \times Q(BG^{ad}) \xrightarrow{\ \wedge\ } Q(BG^{ad} \wedge BG^{ad}) \xrightarrow{\ d^!\ } Q(BG^{ad})$$

and let $\lambda_1: F_G^1(\infty) \to Q(BG^{ad})$ *be the stabilized map* $f \to \lambda_0^{st}(f \mathbin{\backslash\!\backslash} 1)$. *Then the map* $(f, g) \to \lambda_1(fg)$ *is homotopic to the map*

$$(f, g) \to (\lambda_1(f) \otimes \lambda_1(g)) * \lambda_1(f) * \lambda_1(g).$$

In other words, the composition on $F_G^1(\infty)$ is homotopically the "circle operation" associated to the Hopf ring structure on $F_G^0(\infty)$ (compare [8, p. 8]).

PROOF. The previous discussion of this section implies that λ_0^{st} is a map of Hopf rings up to homotopy. Therefore the map $(f, g) \to \lambda_1(f) \otimes \lambda_1(g)$ is homotopic to

$$(f, g) \to \lambda_0^{st}((fg \mathbin{\backslash\!\backslash} 1) * \chi(f \mathbin{\backslash\!\backslash} 1) * \chi(g \mathbin{\backslash\!\backslash} 1)),$$

which in turn is homotopic to

$$(f, g) \to \lambda_1(fg) * \chi\lambda_1(F) * \chi\lambda_1(g),$$

where χ is the loop inverse in $Q(BG^{ad})$. If $\lambda_1(f) * \lambda_1(g)$ is added to the relevant homotopies, one can readily construct a homotopy between the maps specified in the theorem.

4. FINAL REMARKS

As noted in [2, §3], the "identity component" spaces SF_G have canonical infinite loop space structures induced by composition. Likewise, the spaces $F_G^O(\infty)$ admit two E_∞-space structures (in the sense of May [11]) induced by loop sum and composition. The latter structures satisfy all the conditions for E_∞ ring spaces as defined by May (e.g., see [12]) except existence of a multiplicative identity. By Theorem 3.4, composition in F_G is essentially the circle operation associated with this ring structure. Motivated by this example one is led to ask the following general question:

(4.1) PROBLEM. Given an "E_∞ ring space without unit" (i.e., a multiplicative identity does not necessarily exist), does the associated circle operation always extend to an E_∞-space structure?

Theorem 3.4 can be used to give a fairly good description of the Pontrjagin product in $H_*(F_G; Z/p)$, where p is prime; since F_G is homotopic to $Q(BG^{ad})$, the homology of F_G is given by results of Dyer, Lashof, and May [5, 10-12]. Presumably a more thorough study would give complete information on the Pontrjagin product and allow computation of $H_*(BSF_G; Z/p)$ via the Eilenberg-Moore spectral sequence, in analogy with the case $G = 1$ [10-13, 17]. To simplify the discussion we shall assume G is finite abelian (hence cyclic); in this case $BG^{ad} = BG^+$. Henceforth all homology is understood to have Z/p coefficients.

By Theorem 3.4, the calculation quickly reduces to determining the multiplication on $H_*(Q(BG^+))$ induced by $\otimes$. Specifically, if $\textcircled{T}$ denotes the latter multiplication and $\circ$ denotes the operation corresponding to composition in F_G, it is immediate that

$$(4.2) \qquad a \circ b = \sum_{i,j} (-1)^{|a_i''||b_j'|} (a_i' \textcircled{T} b_j') * a_i'' * b_j'',$$

where $*$ is the loop operation, $|\ldots|$ is the dimension of an element, and the diagonal maps send a and b to $\Sigma\, a_i' \otimes a_i''$ and $\Sigma\, b_j' \otimes b_j''$ respectively. Calculation of $\textcircled{T}$ may be done by noticing that the covering $G \to BG \to BG \times BG$ is replaceable by the more manageable map $G \to EG \times BG \to BG \times BG$; to be precise, the former covering pulls back to the latter under the map induced by the shear automorphism $E_{12}(1)$ sending $(x, y) \in G \times G$ to $(x + y, y)$. If $E_{12}(-1)$ denotes the inverse automorphism, then the naturality and multiplicative properties of the transfer imply that $\otimes$ is given by the following threefold composition:

$$
(4.3) \qquad
\begin{array}{ccc}
Q(BG^+) \times Q(BG^+) & \xrightarrow[\text{smash}]{\text{external}} & Q(BG^+ \wedge BG^+) \\
 & & \downarrow{\scriptstyle Q(BE_{12}(-1)^+)} \\
\otimes & & Q(BG^+ \wedge BG^+) \\
 & & \downarrow{\scriptstyle \Omega^\infty(BG^+ \wedge \text{transfe}} \\
 & \longrightarrow & Q(BG^+ \wedge S^0) = Q(BG^+)
\end{array}
$$

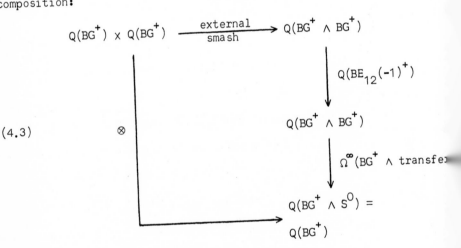

Of course, the map $Q(BE_{12}(-1)^+)_*$ is simple to compute in terms of Dyer-Lashof operations. Similarly, enough is known about the transfer map $BG^+ \to QS^0$ to make the map $\Omega^\infty(BG^+ \wedge \text{transfer})_*$ relatively understandable. Finally, there is the problem of calculating the homology pairing induced by external smash product; to be fully effective, this formula should express the image of a pair $(Q^I a, Q^J b)$, with Q^I, Q^J sequences of Dyer-Lashof operations and $a, b \in \tilde{H}_*(BG^+)$, in terms of Dyer-Lashof operations on classes in $\tilde{H}_*(BG^+ \wedge BG^+)$. Formulas of the desired type for external smash products $Q(X) \times Q(S^0) \to Q(X)$ have been given by May (e.g., [10, Thm. 3.1(iii)]), and I have been informed that similar but more complicated formulas exist for the general case $Q(X) \times Q(Y) \to Q(X \wedge Y)$. Thus a reader who has a special interest in the Pontrjagin ring of $H_*(F_G)$ should be able to make fairly extensive (perhaps even complete) calculations.

REFERENCES

1. J. C. Becker and R. E. Schultz, Equivariant function spaces and stable homotopy theory I, Comment. Math. Helv. 49 (1974), 1-34.
2. J. C. Becker and R. E. Schultz, Equivariant function spaces and stable homotopy theory II, Indian Univ. Math. J. 25 (1976), 481-492.
3. C. Chevalley, Fundamental Concepts of Algebra, Pure and Applied Mathematics Vol. 7. Academic Press, New York, 1956.
4. T. tom Dieck, The Burnside ring and equivariant stable homotopy (notes by M. C. Bix), mimeographed, University of Chicago, 1975.

5. E. Dryer and R. Lashof, Homology of iterated loop spaces, Amer. J. Math. 84 (1962), 35-88.

6. H. Hauschild, Allgemein Lage und Äquivariante Homotopie, Math. Z. 143 (1975), 155-164.

7. H. Hauschild, Äquivariante Homotopie I, Math. Z., to appear.

8. N. Jacobson, Structure of Rings (Rev. Ed.), A.M.S. Colloquium Publications Vol. 37. American Mathematical Society, Providence, 1964.

9. C. Kosniowski, Equivariant cohomology and stable cohomotopy, Math. Ann. 210 (1974), 83-104.

10. J. P. May, Homology operations or infinite loop spaces, Algebraic Topology, Proc. Symposia in Pure Math. 22, 171-186. American Mathematical Society, Providence, 1971.

11. J. P. May, The Geometry of Iterated Loop Spaces, Lecture Notes in Mathematics Vol. 271. Springer, New York, 1972.

12. J. P. May, (with contributions by F. Quinn and N. Ray), E_∞ ring spaces and E_∞ ring spectra, mimeographed, University of Chicago, 1975.

13. R. J. Milgram, The mod 2 spherical characteristic classes, Ann. of Math. 92 (1970), 238-261.

14. R. Schultz, Homotopy decompositions of equivariant function spaces I, Math. Z. 131 (1973), 49-75.

15. R. Schultz, Homotopy decompositions of equivariant function spaces II, Math. Z. 132 (1973), 69-90.

16. G. B. Segal, Equivariant stable homotopy theory, Actes, Congrès internat. math. (Nice, 1970), T. 2, 59-63. Gauthier-Villars, Paris, 1971.

17. A. Tsuchiya, Characteristic classes for spherical fiber spaces, Nagoya Math. J. 43 (1971), 1-39.

18. E. C. Zeeman, A proof of the comparison theorem for spectral sequences, Proc. Camb. Philos. Soc. 53 (1957), 57-62.

PURDUE UNIVERSITY,
WEST LAFAYETTE, 47907,
U.S.A.

A PROPERTY OF A CHARACTERISTIC CLASS OF AN ORBIT FOLIATION

HARUO SUZUKI

INTRODUCTION

We shall present here a vanishing theorem of Bott's exotic characteristic classes for a foliated principal GL_r-bundle with a transverse projectable connection, and its application to the foliations defined by smooth actions of Lie group and its subgroup. By "smooth", we mean "C^∞-differentiable".

Let $\mathcal{F}$ and $\mathcal{F}'$ be smooth foliations of codimensions q and r respectively on a paracompact Hausdorff smooth manifold M. Let F and F′ be tangent subbundles corresponding to the foliations $\mathcal{F}$ and $\mathcal{F}'$ respectively. Suppose that F is a vector sub-bundle of F′. We denote by $\Delta_r: H^*(WO_r) \to H^*_{DR}(M; C)$, the Bott's characteristic homomorphism, where C is the complex number field. We shall prove,

THEOREM 3.1. <u>Let $\mathcal{F}$ and $\mathcal{F}'$ be foliations on</u> M <u>stated in the above. If $\mathcal{F}$ is a Riemannian foliation and</u> $[q/2] \leq r$, <u>then the characteristic classes</u>

$$\Delta_r[c_{i_1} \cdots c_{i_\lambda} \otimes h_{j_1} \wedge \cdots \wedge h_{j_\mu}],$$

$$1 \leq i_1 \leq \cdots \leq i_\lambda \leq r,$$

$$1 \leq j_1 < \cdots < j_\mu \leq r, \qquad (j_\beta \text{ odd}),$$

$$\sum_{\alpha=1}^{\lambda} i_\alpha + j_1 > r + 1$$

are zero. If $q \leq 2r$, then the last condition is reduced just to the cocycle condition, $\sum_{\alpha=1}^{\lambda} i_\alpha + j_1 > r$.

As an application of the above theorem, one can show the following. Suppose that a connected Lie group G acts smoothly on the manifold M and all orbits are of the same codimension r, i. e., the action defines a smooth foliation $\mathcal{F}_G$ of codimension r, on M. If one of Bott's characteristic classes of Theorem 3. 1, is not zero, then no connected subgroup H of G determines a Riemannian foliation $\mathcal{F}_H$ of codimension q ($[q/2] \leq r$, $q \geq r$), under the action of H as subgroup of G. In particular, from the last part of the theorem one obtains,

COROLLARY 3. 3. If $2r \geq q \geq r$ and if one of Bott's exotic characteristic classes of the foliation $\mathcal{F}_G$ is not zero, then no compact connected subgroup H of g has a non-singular orbit foliation on M of codimension q.

In section 1, we give a brief review of Bott's characteristic classes in rather general situation. In section 2, assuming that $\mathcal{F}$ is Riemannian, two transverse projectable connections are constructed on the foliated principal GL_r-bundle $E(M, p_T, GL_r)$ over $(M,\mathcal{F})$, which is the frame bundle $E(V'')$ of the quotient vector bundle $V'' = T(M)/F'$, equipped with a lifted foliation by a Bott's connection of V''. One of them is a Riemannian connection and the other is transverse for the foliated principal GL_r-bundle structure of $E(V'')$, equipped with a lifted foliation over $(M,\mathcal{F}')$ by the Bott's connection. In the last section, the vanishing theorem of Bott's exotic characteristic classes is proved and its application to Lie group actions is given.

1. PRELIMINARIES

Let $\mathcal{F}$ be a smooth foliation of codimension q, on a paracompact Hausdorff smooth manifold M of dimension n and let F be the subbundle of the tangent bundle T(M) of M, which consists of tangent vectors to the leaves of $\mathcal{F}$. We denote the general linear group of order r by GL_r. <u>A foliated principal</u> GL_r-<u>bundle</u> $E(M, p, GL_r)$ over the foliated manifold $(M,\mathcal{F})$ is a smooth principal GL_r-bundle $p: E \to M$, equipped with a right GL_r-invariant smooth foliation $\mathcal{F}_E$ on E, each leaf of which covers a leaf of $\mathcal{F}$. A <u>transverse connection</u> Θ on $E(M, p, GL_r)$ over $(M, \mathcal{F})$ is a connection Θ on E, for which leaves of the lifted foliation $\mathcal{F}_E$ are horizontal (cf. [2, Definition 2.1] and [3]).

A Bott's connection [1, Definition (6. 1)] on the quotient Vector bundle $V = T(M)/F$ makes the frame bundle E_T of V a foliated principal GL_r-bundle over $(M,\mathcal{F})$, where the Bott's connection is transverse. It is denoted by $E_T(M, p_T, GL_q)$.

A transverse connection ω on $E(M, p, GL_r)$ over $(M,\mathcal{F})$ is called <u>projectable</u> if it is locally induced from a connection of the restriction of $E(M, p, GL_r)$ on a transverse q dimensional submanifold to $\mathcal{F}$, by the local projection of M along leaves of $\mathcal{F}$ (cf. [2, 2. 24], [3] or [4]).

Let $WO_{q,r}$ be the differential algebra;

$$R[c_1, \dots, c_s]/(\deg > q) \otimes \Lambda(h_1, \dots h_3, \dots, h_\ell),$$

where R is the real number field, $s = \min(q, r)$, $\ell = \max\{2m + 1 \leq r\}$ and the differential is given by $d(c_i) = 0$, $dh_j = c_j$ for $j \leq s$ and $dh_j = 0$ for $j > s$ if $q < \ell$. Taking the curvature form of a transverse connection Θ on $E(M, p, GL_r)$ over $(M,\mathcal{F})$ and that of the affine combination of Θ and a Riemannian connection Θ^0 on (EM, p, GL_r), the homomorphism $\Delta_{q,r}$: $H^*(WO_{q,r}) \to H^*_{DR}(M; C)$ of Bott's characteristic classes of $E(M, p, GL_r)$ over $(M,\mathcal{F})$ is defined, where C is the complex number field. $\Delta_{q,r}$ does not depend on a choice of Θ (and of Θ^0) (c.f. [7, Proposition 1. 5]).

If the foliated principal GL_r-bundle $E(M, p, GL_r)$ over $(M,\mathcal{F})$ admits a transverse projectable connection ω, we have a characteristic homomorphism $\Delta_{[q/2],r}(\omega)$: $H^*(WO_{[q/2],r}) \to H^*_{DR}(M; C)$ in a similar way to $\Delta_{q,r}$. Let T: $WO_{q,r} \to WO_{[q/2],r}$ denote the cochain map defined by the natural projection homomorphism, then we have a commutative diagram,

$$\begin{array}{ccc} H^*(WO_{q,r}) & \xrightarrow{\Delta_{q,r}} & \\ \downarrow T^* & & H^*_{DR}(M; C) \\ H^*(WO_{[q/2],r}) & \xrightarrow{\Delta_{[q/2],r}(\omega)} & \end{array}$$

THEOREM 1. 1. ([7, Proposition 2. 2]) *Let* $E(M, p, GL_r)$ *be a foliated principal* GL_r-*bundle over a foliated manifold* $(M,\mathcal{F})$, *with codimension* q *foliation* $\mathcal{F}(q \geq r)$, *which admits a transverse projectable connection* ω. *Let* γ *be a cocycle of the differential complex* $WO_{[q/2],r}$, *given by*

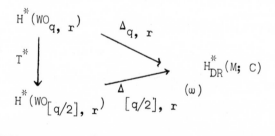

$$\gamma = c_{i_1} \cdots c_{j_\alpha} \otimes h_{j_1} \wedge \cdots \wedge h_{j_\mu},$$

$$1 \leq i_1, \ldots, i_\lambda \leq r,$$

$$1 \leq j_1 < \ldots < j_\mu \leq \ell \qquad (j_\beta \text{ odd}),$$

$$2(\sum_{\alpha=1} i_\alpha + j_1) > q + 1.$$

Then $\Delta_{[q/2], r}(\omega)[\gamma]$ does not depend on a choice of ω, where $[\gamma]$ denote the cohomology class of γ. In particular, if q is even, then the last condition is reduced just to the cocycle condition $2(\sum_{\alpha=1}^{\lambda} i_\alpha + j_1) > q$, that is, the homomorphism $\Delta_{[q/2], r}(\omega)$ does not depend on a choice of ω.

REMARK. Let $g\ell(r)$ denote the Lie algebra of $GL(n, R)$. The cohomology ring $H^*(W(g\ell(r), O(r))_q)$ of [2] is precisely $H^*(WO_{q, r})$.

2. CONNECTIONS IN VECTOR BUNDLES

Let V be a smooth vector bundle of dimension q, over a paracompact Hausdorff smooth manifold M and $E(V)$, the principal GL_q-bundle associated ot V, that is, the frame bundle of V. It is well known that a connection on $E(V)$ determines a unique connection of covariant derivative on the vector bundle V and the converse also holds. We identify the both connection on V and E.

LEMMA 2. 1. Let V' be an r dimensional smooth vector subbundle of V. A connection ∇ on V induces a connection ∇' on the subbundle V'. Moreover, if ∇ is Riemannian then so is ∇'.

PROOF. V splits smoothly into Whitney sum $V = V' \oplus V''$, where V'' is a smooth $q - r$ dimensional vector subbundle of V. Let $\sigma: V \to V'$ be the projection map determined by this splitting and $i: V' \to V$, the inclusion map. Let $\Gamma(.)$ denote a module of smooth sections of a smooth vector bundle. One defines an R-bilinear map

$$\nabla': \Gamma(T(M)) \times \Gamma(V') \to \Gamma(V')$$

by $\nabla'(X, s) = \sigma\nabla(x, i(s))$, and denote it by $\nabla'_X(s)$ for all $X \in \Gamma(T(M))$ and $s \in \Gamma(V')$, It is easy to see that ∇' is a covariant derivative and determines a connection on V'.

By taking a local orthonormal smooth frame field of V', the Riemannian condition for ∇' follows from that for ∇. Q. E. D.

Let N be a smooth manifold and $h: N \to M$ a smooth map. In the induced vector bundle h^*V, we take a smooth local frame section $(d_1, \dots d_q)$ such that

$$(1) \qquad d_j(y) = (e_j(h(y)),y) \in V \times N$$

for $y \in N$, where $(e_1, \dots, e_q)$ is a smooth local frame section of V, defined on a neighborhood U of $h(y)$. d_j is defined on the neighborhood $\overline{U} = h^{-1}(U)$ of y. For d_j, $j=1, \dots ,q$ and for $Y \in \Gamma(T(N)) \mid_{\overline{U}}$ we define $(h^*\nabla)_Y$ by

$$(2) \qquad ((h^*\nabla)_Y(d_j))(y) = (\nabla_{h_*Y(y)}(e_j), y)$$

and extend it to the set of smooth local section of h^*V under the conditions of covariant derivative. $(h^*\nabla)_Y$ does not depend on a choice of $(d_1, \dots, d_q)$ and hence $(h^*\nabla)_Y(s)$ is well defined for all $Y \in \Gamma(T(N))$ and all $s \in \Gamma(h^*V)$.

$h^*\nabla: \Gamma(T(N)) \times \Gamma(h^*V) \to \Gamma(h^*V)$ is an induced connection of ∇ by the map h, on h^*V.

LEMMA 2. 2. The induced connection by the smooth map h is compatible with the construction of connection for a vector subbundle given by Lemma 2. 1, that is,

$$h^*\nabla' = (h^*\nabla)'.$$

PROOF. We have a splitting of h^*V into the Whitney sum

$$h^*V = h^*V' \oplus h^*V'',$$

determined by the splitting $V = V' \oplus V''$. The projection map $h^*V \to h^*V'$, induced by σ is denoted also by σ. Let $i: h^*V' \to h^*V$ denote also the inclusion map. For local smooth frame section $(d_1, \dots d_r)$ of h^*V', satisfying the condition corresponding to (1), we have,

$$\begin{aligned}(i.d_j)(y) &= i(e_j(h(y)), y) \\ &= ((i.e_j)(h(y)), y).\end{aligned}$$

By definition (2) of $h^*\nabla'$, and using the above relation, one obtains,

$$\begin{aligned}((h^*\nabla')_Y(d_j))(y) &= (\nabla'_{h_*Y(y)}(e_j), y) \\ &= \sigma(\nabla_{h_*Y(y)}(i.e_j), y) \\ &= ((h^*\nabla')_Y(d_j))(y),\end{aligned}$$

for $Y \in \Gamma(T(N))|_{\bar{U}}$, $\bar{U}$ being a neighborhood of y. Extending this relation to the set of smooth local sections of h^*V' under the conditions of covariant derivative and taking account of its independence of a choice of local frame $(d_1, \dots d_r)$, we have the conclusion, $h^*\nabla' = (h^*\nabla)'$. Q. E. D.

COROLLARY 2. 3. *Let* V *be a smooth vector bundle over* M *and* V', *a smooth vector subbundle of* V. *A connection* ∇ *on* V *induces a connection* ∇'' *on* $V'' \cong V/V'$ *and the induced connection by the smooth map* h *is compatible with the construction of* ∇'', *that is,*

$$h^*\nabla'' = (h^*\nabla)''.$$

We denote frame bundles of V'' and h^*V'' by $E(V'')$ and $E(h^*V'')$ respectively. Corollary 2. 3 means that the connection on $E(h^*V'')$ obtained from the induced connection $h^*\nabla$ on $E(h^*V)$ is the connection induced from ∇'' on $E(V'')$ by h.

Let $\mathcal{F}$ and $\mathcal{F}'$ be smooth foliations of codimensions q and r respectively on the smooth manifold M. Let F and F' be tangent subbundles of $T(M)$, corresponding to $\mathcal{F}$ and $\mathcal{F}'$ respectively. Suppose that F is a subbundle of F'. We denote by $E(M, p_T, GL_q, \mathcal{F}_{E,\,q})$, the frame bundle $E(V)$ of the quotient bundle $V = T(M)/F$ which is equipped with a foliated principal GL_q-bundle structure by a lifted foliation $\mathcal{F}_{E,\,q}$ over $(M,\mathcal{F})$. And also we denote by $E(M, p_T, GL_r, \mathcal{F}_{E,\,r})$, the frame bundle $E(V'')$ of the quotient bundle $V'' = T(M)/F'$ which is equipped with a foliated principal GL_r-bundle structure by a lifted foliation $\mathcal{F}_{E,\,r}$ over $(M,\mathcal{F})$. We apply Corollary 2. 3 to the quotient bundle $V'' = T(M)/F' \cong V/V'$ and then obtain the following.

THEOREM 2. 4. *If the foliated principal* GL_q*-bundle* $E(M, p_T, GL_q, \mathcal{F}_{E,\,q})$ *over* $(M,\mathcal{F})$ *has a transverse projectable connection* ∇, *then there is a foliated principal* GL_r*-bundle* $E(M, p_T, GL_r, \mathcal{F}_{E,\,r})$ *over* $(M,\mathcal{F})$ *which admits also a transverse projectable connection* ∇''. *Moreover, if* ∇ *is Riemannian then so is* ∇''.

PROOF. V splits smoothly into the Whitney sum $V = V' \oplus V''$. Let ∇ denote the transverse projectable connection on $E(M, p_T, GL_q, \mathcal{F}_{E,q})$ over $(M,\mathcal{F})$. It is also regarded as a connection on V locally induced from a connection $\bar{\nabla}$ on a restriction of V over a q dimensional smooth local submanifold N transverse to $\mathcal{F}$ in M, by the local projection h to N along leaves of $\mathcal{F}$. One can assume that N is diffeomorphic to the q dimensional Euclidean space R^q.

By applying Corollary 2. 3 to the map h, one obtains a connection ∇'' on V'', which is locally induced from a connection $\bar{\nabla}''$ on the restriction bundle $V''|N$ by h. Because the connection form of ∇'' vanishes along leaves of a local foliation on $E(V'')$ which is defined by the local submersion covering h, $E(V'')$ admits a right GL_r-invariant foliation $\mathcal{F}_{E,r}$ lifting $\mathcal{F}$ and ∇'' is a transverse projectable connection on the foliated principal bundle $E(M, p_T, GL_r, \mathcal{F}_{E,r})$ over $(M,\mathcal{F})$.

If ∇ is Riemannian then so is ∇'' by the last part of Lemma 2. 1. Q. E. D.

A foliation $\mathcal{F}$ on M is said to be <u>Riemannian</u> if transition functions of local submersions of $\mathcal{F}$ are isometries. Now we have the following.

THEOREM 2. 5. *Let $\mathcal{F}$ and $\mathcal{F}'$ be smooth foliations on M such that the tangent subbundle F to $\mathcal{F}$ is a subbundle of the tangent subbundle F′ to $\mathcal{F}'$. Suppose that the lifted foliation $\mathcal{F}_{E,r}$ of $E(M, p_T, GL_R, \mathcal{F}_{E,r})$ over $(M,\mathcal{F})$ is determined by the Bott's connection of $V'' = T(M)/F'$. If $\mathcal{F}$ is Riemannian, then there is a* transverse projectable connection on $E(M, p_T, GL_r, \mathcal{F}_{E, r})$ over $(M,\mathcal{F})$, which is also a transverse connection on $E_T(M, p_T, GL_r)$ over $(M,\mathcal{F}')$.

PROOF. Let $\{U_i\}$ be an open cover of M, by coordinate neighborhoods of local submersions f_i for $\mathcal{F}$ and W, the disjoint union of $f_i U_i (\subset R^q)$. $\mathcal{F}'$ induces a smooth foliation $\bar{\mathcal{F}}$ of codimension r on W which is invariant under the transition function γ_{ij} of local submersions for $\mathcal{F}$. Since γ_{ij} is an isometry by assumption, one can find a Bott's connection of W, which is invariant under γ_{ij}. Therefore, this induces a transverse projectable connection on the $E(M, p_T, GL_r, \mathcal{F}_{E, r})$ over $(M,\mathcal{F})$. It is easy to see that leaves of the lifted foliation $\mathcal{F}'_{E_T}$ on $E_T(M, p_T, GL_r)$ over $(M,\mathcal{F}')$ by the Bott's connection, are horizontal with respect to this connection. Q. E. D.

3. PROOF OF MAIN THEOREMS

Let $\mathcal{F}$ and $\mathcal{F}'$ be smooth foliations of codimensions q and r respectively on a paracompact Hausdorff smooth manifold M. Let F and F′ be tangent subbundles of T(M) corresponding to $\mathcal{F}$ and $\mathcal{F}'$ respectively. Suppose that F is a vector bundle of F′. We describe this condition simply by $\mathcal{F} \subset \mathcal{F}'$.

We denote the differential complex $WO_{r, r}$ by WO_r and the Bott's characteristic homomorphism $\Delta_{r, r}: H^*(WO_r) \to H^*_{DR}(M; C)$ by Δ_r. Now we prove a vanishing theorem of Bott's exotic characteristic classes.

THEOREM 3. 1. Let $\mathcal{F}$ and $\mathcal{F}'$ be smooth foliations of codimensions q and r respectively, on M, such that $\mathcal{F} \subset \mathcal{F}'$. If

$\mathcal{F}$ is Riemannian and $[\frac{q}{2}] \leq r$, then the characteristic classes

$$\Delta_r[c_{i_1} \cdots c_{i_\lambda} \otimes h_{j_1} \wedge \cdots \wedge h_{j_\mu}],$$

$$1 \leq i_1, \ldots, i_\lambda \leq r,$$

$$1 \leq j_1 < \ldots < j_\mu \leq r \qquad (j_\beta \text{ odd}),$$

$$\sum_{\alpha=1}^{\lambda} i_\alpha + j_1 > r + 1$$

are zero, where $[\cdot]$ means a cohomology class. If $q \leq 2r$ then the last condition is reduced just to the cocycle condition $\sum_{\alpha=1}^{\lambda} i_\alpha + j_1 > r$.

PROOF. Let $E_T(M, p_T, GL_r)$ be the foliated principal GL_r-bundle over $(M, \mathcal{F}')$, determined by a Bott's connection Θ of $V'' = T(M)/F'$. We note that $E_T(M, p_T, GL_r)$ induces a foliated principal GL_r-bundle $E(M, p_T, GL_r, \mathcal{F}_{E,r})$ over $(M, \mathcal{F})$ on which Θ is also transverse. Since $\mathcal{F}$ is Riemannian, by Theorem 2. 5, we have a transverse projectable connection ω on $E(M, p_T, GL_r, \mathcal{F}_{E,r})$ over $(M, \mathcal{F})$, which is also a transverse connection on $E_T(M, p_T, GL_r)$ over $(M, \mathcal{F}')$. Therefore, we may use ω to define the Bott's characteristic homomorphism Δ_r for the foliation $\mathcal{F}'$.

On the other side, ω determines a characteristic homomorphism $\Delta_{[q/2], r}(\omega): H^*(WO_{[q/2], r}) \to H^*_{DR}(M; C)$ (cf. Section 1). Let $T: WO_r \to WO_{[q/2], r}$ be the cochain map induced by the natural projection, then we have a commutative diagram,

$$(3) \qquad \begin{array}{ccc} H^*(WO_r) & \xrightarrow{\ \Delta_r\ } & \\ \Big\downarrow T^* & & H^*_{DR}(M;\ C) \\ H^*(WO_{[q/2],\, r}) & \xrightarrow{\ \Delta_{[q/2],\, r}(\omega)\ } & \end{array}$$

By Theorem 2. 4, we have also a transverse projectable connection ω^0 on $E(M, p_T, GL_r\ \mathcal{F}_{E,\, r})$ over $(M, \mathcal{F})$, which is Riemannian. And hence one can see that $\Delta_{[q/2],\, r}(\omega^0)(T^*[\gamma]) = 0$ for $\gamma = c_{i_1} \cdots c_{i_\lambda} \otimes h_{j_1} \wedge \cdots \wedge h_{j_\mu}$. If $\sum_{\alpha=1}^{\lambda} i_\alpha + j_1 > r + 1$ or $q \leq 2r$ then from Theorem 1. 1 and the commutativity of (3), it follows that

$$\begin{aligned} \Delta_r[\gamma] &= \Delta_{[q/2],\, r}(\omega)(T^*[\gamma]) \\ &= \Delta_{[q/2],\, r}(\omega^0)(T^*[\gamma]) \\ &= 0. \end{aligned}$$

Q. E. D.

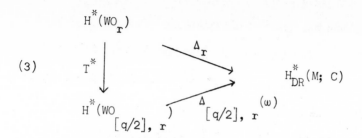

Now we consider orbits of a smooth action of Lie group on a smooth manifold. If all orbits are of the same codimension, then the family of orbits defines a smooth foliation of this codimension, which is called an <u>orbit foliation</u>. Let M be a paracompact Hausdorff smooth manifold and G, a connected Lie group which acts smoothly on M. Let H be a connected subgroup of G. Suppose that orbit of G are of the same codimension r and those of H are the same codimension q. The foliations determined by G and H are denoted by $\mathcal{F}_G$ and $\mathcal{F}_H$ respectively. It is easy to see that $\mathcal{F}_H \subset \mathcal{F}_G$. Therefore, arguments in Section 2 can be applied to the foliations $\mathcal{F}_G$ and $\mathcal{F}_H$, and one obtains,

THEOREM 3. 2. *Let* $\mathcal{F}_G$ *and* $\mathcal{F}_H$ *be orbit foliations of a connected Lie group* G *and its connected subgroup* H *on a paracompact Hausdorff smooth manifold* M *of dimension of* n. *Let* r *and* q $(n \geq q \geq r)$ *be codimensions* $\mathcal{F}_G$ *and* $\mathcal{F}_H$ *respectively. If* $[q/2] \leq r$ *and* H *is compact, then the Bott's characteristic classes of* $\mathcal{F}_G$,

$$\Delta_r[c_{i_1} \cdots c_{i_\lambda} \otimes h_{j_1} \wedge \cdots \wedge h_{j_\mu}],$$

$$1 \leq i_1, \dots, i_\lambda \leq r,$$

$$1 \leq j_1 < \dots < j_\mu \leq r \qquad (j_\beta \text{ odd}),$$

$$\sum_{\alpha=1}^{\lambda} i_\alpha + j_1 > r + 1$$

are zero. If $q \leq 2r$, *then the last condition is reduced just to the cocycle condition* $\sum_{\alpha=1}^{\lambda} i_\alpha + j_1 > r$.

PROOF. Since H is a compact connected Lie group, by [5, §5], M has a bundle-like metric with respect to the foliation $\mathcal{F}_H$, that is, $\mathcal{F}_H$ is a Riemannian foliation. Theorem 3. 1 can be applied to foliations $\mathcal{F}_G \supset \mathcal{F}_H$ and the result follows immediately. Q. E. D.

As a direct consequence of Theorem 3. 2, we have,

COROLLARY 3. 3. *Let the manifold* M *and the Lie group* G *be as in Theorem* 3. 2. *If* $2r \geq q \geq r$ *and if one of Bott's exotic characteristic classes of* $\mathcal{F}_G$ *is not zero, then no compact connected subgroup* H *of* G *has a non-singular orbit foliation on* M, *of codimension* q.

REFERENCES

1 R. Bott, Lecture on characteristic classes and foliations, Lecture Notes in Mathematics 279, Springer, Berlin-New York, 1972, 1-76.

2 F. Kamber and Ph. Tondeur, Foliated bundles and characteristic classes, Lecture Notes in Mathematics 493, Springer, Berlin-New York 1975.

3 P. Molino, Classe d'Attiyah d'un feuilletage et connexions transverses projectables, C.R. Acad. Sci, Paris Ser. A-B 272 (1971), A779-A781.

4 P. Molino, Propriétés cohomologiques et propriétés topologiques des feuilletages a connexion transverse projectable, Topology 12 (1973), 317-325.

5 J.S. Pasternack, Foliations and compact Lie group actions, Comment. Math. Helv. 46 (1971), 467-477.

6 B. Reinhart, Foliated manifolds with bundle-like metrics, Ann. of Math. 69 (1959), 119-132.

7 H. Suzuki, Characteristic classes of foliated principal GL_r-bundles, Hokkaido Math. J. 4 (1975), 159-168.

HOKKAIDO UNIVERSITY,
SAPPORO,
JAPAN.

ORBIT STRUCTURE FOR LIE GROUP ACTIONS ON HIGHER COHOMOLOGY PROJECTIVE SPACES

PER TOMTER

INTRODUCTION

In this paper we describe the cohomological orbit structure of Lie group actions on cohomology projective spaces and show that under quite general conditions an arbitrary action has the same structure as that obtained from an equivariant James' reduced product construction applied to an action of the group on a sphere. Here X is a cohomology projective space if $X \sim P^n(q)$, i.e. $H^*(X) = Q[e]/(e^{n+1})$ as a Q-algebra, where deg $e = q$ is even. Cohomology is always taken with rational coefficients. If $q = 2$, X is a cohomology complex projective space and if $q = 4$, X is a cohomology quaternionic projective space. For those cases linear actions on the classical projective spaces demonstrate that tori of large rank can have rich orbit structures (e.g. many non-acyclic components of the fixed point set). The work of Wu Yi Hsiang [12] and Hsiang and Su [13] shows that the cohomogical orbit structure of an arbitrary action is modelled after the linear examples in general. In particular, there is the following theorem of Hsiang and Su: If $X \sim P^n(4)$ and a torus of rank at least two acts cohomology effectively on X, then at most one component of the fixed point set is a $P^r(4)$ with $r > 0$.

Our work is concerned with the higher cohomology projective

spaces, i.e. $q > 4$. From Corollary 1 in section 2 it follows that in this case the situation is much more rigid, and completely different from the classical complex and quaternionic cases. Moreover, for tori of rank at least 5, there is correspondence between theory and examples, and an action on $X \sim P^n(q)$ can be analyzed in terms of a generating action on S^q, we explain this connection and give some examples of applications in section 3.

We note that if $n = 2$, X is a cohomology projective plane. In [5] Chang and Comenetz show that if $q > 4$ and $\mathrm{rk}\, T > \log_2 q$, then the action is of spherical type, as defined in section 2. Using Skjelbred's application of Sylvester's theorem (see Remark, section 2) it is obviously sufficient to assume $\mathrm{rk}\, T > 4$ for cohomology projective planes. For larger n, however, the fixed point set may have many non-acyclic components in general, and this argument cannot be applied directly. However, Theorem 6, which generalizes the result of Hsiang and Su, allows a similar application of Sylvester's theorem.

In section 2 we define the geometric weight system for actions on $X \sim P^n(q)$ when the rank is larger than 4. This is an invariant from the equivariant cohomology of the space which is simple enough to be effectively computable, on the other hand strong enough to determine the cohomological orbit structure of X. This means the following: The connected orbit types $G/G_x{}^o$ of X are determined by the identity components of the isotropy subgroups, G_X. If $x \in X$, the F^o-variety of x, $F^o(x)$, is the connected component of x in the fixed point set of $G_x{}^o$. The structure of this network of F^o-varieties determines the orbit structure of X. Thus, in particular, the geometric weight system should determine all connected orbit types, and the cohomological structure of the corresponding F^o-varieties.

The basic tool for setting up the geometric weight system is a linearity theorem for certain ideals associated to the equivariant cohomology algebra. This idea goes back to the "topological Schur lemma" of Wu-Yi Hsiang [10]. Special cases of annihilator ideals of submodules of $H_G(X,F)$ were studied in [17]. A quite general structure theory for annihilator ideals of such submodules was developed by T. Chang and T. Skjelbred and has found interesting applications. Their theory is not general enough for our situation, where it is necessary to consider the primary decomposition of annihilators of quotients of submodules of $H_G(X,F)$. Some new considerations are needed here, and in section one, after a few remarks on the basic notions and theorems of equivariant cohomology, we prove the relevant splitting theorem for such ideals. This version is repeatedly applied in the rest of the paper.

1. STRUCTURE THEOREMS IN EQUIVARIANT COHOMOLOGY

Let G be a compact Lie group. All G-spaces X are assumed to be paracompact, of finite cohomological dimension and with a finite number of orbit types. $X \sim Y$ means that $H^*(X)$ is isomorphic to $H^*(Y)$ as a Q-algebra. For standard constructions we refer to Bredon [4] or Hsiang [12]. Thus X_G is the total space of the fibre bundle associated to the universal G-bundle : $E_G \to B_G$ by the given G-action on X. The equivariant cohomology of X is defined by $H^*_G(X) = H^*(X_G)$. If Y is an H-space; $\rho: G \to H$ is a homomorphism of compact Lie groups, and $f: X \to Y$ is ρ-equivariant, there is an induced homomorphism from $H^*_H(Y)$ to $H^*_G(X)$. We need more information on this homomorphism if Y = X and f is the identity. G acts freely on $E_G \times E_H$ by $(e_1, e_2).g = (e_1.e_g, e_2.\rho(g))$; hence we may take $E_G \times E_H$ as the total space in a universal bundle for G. There is a well-defined map: $X_G = (E_G \times E_H) \times_G X \to E_H \times_{\rho(G)} X \to E_H \times_H X = X_H$ given by $(e_1, e_2, x) \to (e_2, x)$. The fibre of this map from X_G to X_H is easily seen to be H_G. When G is connected, the classifying space B_G is simply connected. The Eilenberg-Moore spectral sequence is a 2 quadrant spectral sequence (E_r, d_r) where $E_r \Rightarrow E_\infty = H^*_G(X)$ and $E_2 = \mathrm{Tor}_{RH}(RG, H^*_H(X))$. Here we denote $H^*(B_G)$ by RG; RG and $H^*_H(X)$ are RH-modules through cup-product and the homomorphisms induced in cohomology from the commutative diagram of fibrations:

$$\begin{array}{ccccc} & & X & \longrightarrow & X \\ & & \downarrow & & \downarrow \\ H_G & \longrightarrow & X_G & \longrightarrow & X_H \\ \downarrow & & \downarrow & & \downarrow \\ H_G & \longrightarrow & B_G & \longrightarrow & B_H \end{array}$$

If RG or $H^*_H(X)$ is a flat RH-module, it is well known that $\mathrm{Tor}^n_{RH}(RG, H^*_H(X)) = 0$ for $n \neq 0$ and $E_2 = \mathrm{Tor}^0_{RH}(RG, H^*_H(X)) = RG \otimes_{RH} H^*_H(X)$. Hence we have the following result:

THEOREM 1. *If* RG *or* $H^*_H(X)$ *is a flat RH-module, the above Eilenberg-Moore spectral sequence collapses and* $H^*_G(X) = H^*_H(X) \otimes_{RH} RG$;

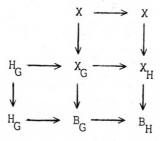

i.e. $H_G^*(X)$ is obtained from $H_H^*(X)$ by an extension of scalars corresponding to the canonical homomorphism $\rho^* : RH \to RG$.

The assumptions of the Theorem are satisfied in the following special cases:

a) G is a subgroup of H and X is totally non-homologous to zero in the fibration $X \to X_H \to B_H$. Then $H_G^*(X) = H_H^*(X) \otimes_{RH} RG$. If G = (e) is the trivial subgroup, we get $H^*(X) = H_H^*(X) \otimes_{RG} Q$.

b) G is a torus, K is a subtorus, and ρ is the epimorphism $G \to H = G/K$. Then $H_G^*(X) = H_{G/K}^*(X) \otimes_{R(G/K)} RG$.

c) G is a maximal torus in the compact, connected Lie group H. Then $H_G^*(X) = H_H^*(X) \otimes_{RH} RG$, and $H_H^*(X) = H_G^*(X)^W$ where W is the Weyl group.

PROOF. In case a) it is obvious from the Serre spectral sequence of $X \to X_H \to B_H$ that $H_H^*(X)$ is a free RH-module; hence it is flat. In case b) it is easy to see that the fibre $H_G = E_G \times_G (G/K) \simeq B_K$ We may identify RG with the polynomial algebra $Q[t_1, \ldots, t_r]$ where the t_i's are identified with linear functionals on G; i.e. elements of $H^1(G)$, via transgression in the universal bundle $G \to E_G \to B_G$. It is then obvious that RG is a free R(G/K)-module. For c) we notice that in general, if G is a subgroup of H, then E_H is also an E_G and there is a fibration from $H_G = E_H \times_G H$ to H_H with fibre H/G; since $H_H \simeq B_{(e)}$ is acyclic, it follows from the Serre spectral sequence that $H_H^*(H_G) \simeq H^*(H/G)$. Let G be a maximal torus in H, let N(G) be the normalizer of G in H and W = N(G)/G the Weyl group. Then H/N(G) is Q-acyclic and the Serre spectral sequence of the fibration $H/N(G) \to X_{N(G)} \to X_H$ shows that $H_H^*(X) = H_G^*(X)^W$, $RH = RG^W$.

Clearly $RG = RH \otimes_Q H^*(H/G)$ is a free RH-module, hence it is flat, and the proof of Theorem 1 is complete.

Now if $x \in X$, let r_x be the canonical projection from RG to RG_x induced by inclusion of G_x in G. If S is a multiplicative subset of RG, let $X^S = \{x \in X;\ S \cap \ker(r_x) = \emptyset\}$. The basic localization theorem for equivariant cohomology is now well known.

THEOREM 2. The localized restriction homomorphism $S^{-1}H_G^*(X) \to S^{-1}H_G^*(X^S)$ is an isomorphism.

If S is the complement of a prime ideal P, we denote $S^{-1}H_G^*(X)$ by $H_G^*(X)_P$ and X^S by X^P. If $P = (0)$, $X^P = X^G = F$ is the fixed point set, and $H_G^*(X)_{(0)} = H_G^*(X) \otimes_{RG} R'G = H_G^*(F) \otimes_{RG} RG' = (H^*(F) \otimes_Q RG) \otimes_{RG} R'G = H^*(F) \otimes_Q R'G$, where R'G is the quotient field of RG.

From now on we assume that $G = T$ is a torus. There are examples of Hsiang which show that only in this case is there a strong relationship between the algebraic structure of the equivariant cohomology and the orbit structure of X. Let $\{x_i\}$ and $\{v_j\}$ be a set of even - and odd-dimensional generators of $H_T^*(X)_{(0)}$, respectively. Then there is a presentation of $H_T^*(X)_{(0)}$ given by an epimorphism p from the free, anti-commutative R'T-algebra $A_T = R'T[x_1,\ldots,x_k] \otimes_{R'T} \Lambda_{R'T}[v_1,\ldots,v_\ell]$ to $H_T^*(X)_{(0)}$. Let p_j: $H_T^*(F)_{(0)} \to H_T^*(F^j)_{(0)}$ be induced from the inclusion of the j-th component F^j into F, let $I = \ker p$ and $I_j = \ker(p_j\ p)$.

THEOREM 3. (Hsiang [11]). 1. The radical of I is the intersection of s maximal ideals M_j whose varieties are rational points $a_j = (a_{j1},\ldots,a_{jk}) \in (R'T)^k$; $i = 1,\ldots,s$.

2. There is a natural bijection between the connected components F^j of F and the above points $\{a_j\}$, such that the

restriction homomorphism of an arbitrary point $q_j \in F^j \subseteq X$ maps the even generator $x_i \in H_T^*(X)_{(0)}$ to $a_{ji} \in H_T^*(\{q_j\})_{(0)} \simeq R'T$.

3. $H^*(F^j) \otimes_Q R'T \simeq A_T/I_j$, where $I_j = I_{M_j} \cap A_T$. Moreover $I = I_1 \cap \ldots \cap I_s = I_1 \ldots I_s$.

Let X be a cohomology manifold over Q; then any component F^j of F is also a cohomology manifold over Q. Let $w_i \in H^2(B_T)$ and let $H_i = (w_i^{\perp})$ be the corank one subtorus whose Lie algebra is the kernel of w_i interpreted as a linear functional. Let $X^{H_i} = G_i^1 + \ldots + G_i^{\ell}$ with the G_i^k's connected; then each F^j is included in a unique $G_i^{i(j)}$. w_i is a local geometric weight at F^j if $\dim G_i^{i(j)} - \dim F^j > 0$, and the multiplicity is defined to be $\frac{1}{2}(\dim G_i^{i(j)} - \dim F^j)$. The local Borel formula asserts that the G_i^k's are transversal in the sense that $\dim X - \dim F^j = \sum_i (\dim G_i^{i(j)} - \dim F^j)$. Let $x \in X$ and $F^j \subseteq F^o(x)$; let $\{w_k, m_k\}$ be the local geometric weight system at F^j. Then $G_x^o = (\cap H_k;\ H_k = (w_k^{\perp}) \supseteq G_x^o)^o$, and $\dim F^o(x) - \dim F^j = 2\Sigma m_k$ (sum over the k's such that $H_k \supseteq G_x^o$). This reveals the significance of the local geometric weight system.

After the proof of the Su conjecture this can be generalised to Poincaré duality spaces over Q (see Chang and Skjelbred [6,8]). A torus $L \subseteq T$ is said to be cohomology ineffective on X if $H^*(X, X^L) = 0$. T acts cohomology effectively if the only cohomology ineffective subtorus is the trivial subgroup. An F^o-variety in X with generic isotropy subgroup $K = K^o$ is then a component V of X^K such that the action of T/K on V is cohomology effective. Then the above statements hold in the more general setting of Poincaré duality spaces over Q when dimension is now interpreted as formal dimension. If X is a compact, orientable cohomology manifold, the two notions of local geometric weights coincide.

We will use the following observation: Let K be a subtorus of T and let P_K be the kernel of the homomorphism $r_K\colon RT \to RK$. The variety of the ideal P_K is the Lie algebra of K; this determines a bijective correspondence between subtori of T and linear subspaces of the Lie algebra of T which are rational with respect to the Q-structure determined by the defining lattice of the torus P. It follows that to a given prime ideal P in RT there exists a unique minimal subtorus K in T such that $P_K \subseteq P$, hence $X^P = X^{P_K} = X^K$.

Let X be a T-space with $F = X^T \neq \emptyset$ and K a subtorus of T. Let M be a submodule of $H^*(F)$ and define $M_{T,K} = \partial(M \otimes RT) \subseteq H^*_T(X^K, F)$, where ∂ is the boundary operator in the exact sequence in the equivariant cohomology for the pair (X^K, F). If K is the trivial subgroup (e), we denote $M_{T,K}$ by M_T simply. Let ρ be the projection from T to $K' = T/K$. It follows from Theorem 1 that $H^*_T(X) \simeq H^*_{K'}(X) \otimes_{RK'} RT$, similarly for F, so $H^*_T(X;F) \simeq H^*_{K'}(X,F) \otimes_{RK'} RT$.

THEOREM 4. *Let X be a T-space with $F = X^T \neq \emptyset$. Let M and N be submodules of $H^*(F)$ with $N \subset M$. Then the prime ideals corresponding to a reduced primary decomposition of $\mathrm{Ann}(M_T/N_T) = \{a \in RT;\ a M_T \subseteq N_T\}$ are linear ideals. The isolated primes $P_1, \ldots, P_\ell$ are characterized as follows: A prime ideal P of RT is equal to one of the P_i, $i = 1, \ldots, \ell$ if and only if $P = P_K$, where K is a maximal subtorus of T with respect to the property $M_{T,K} \neq N_{T,K}$.*

We need a lemma for the proof.

LEMMA 1. *Let K be a subtorus of T. Then all primary ideals associated with a reduced primary decomposition of $\mathrm{Ann}(M_{T,K}/N_{T,K})$*

are contained in P_K.

PROOF. RT is a flat RK'-module. (K' = T/K); hence it is easily seen that $M_{T,K} = M_{K',K} \otimes_{RK'} RT$ and $M_{T,K}/N_{T,K} \simeq (M_{K',K}/N_{K',K}) \otimes_{RK'} RT$. It is well known that in the flat case we must then have $\mathrm{Ann}(M_{T,K}/N_{T,K}) = \mathrm{Ann}(M_{K',K}/N_{K',K}) \otimes_{RK'} RT$. The generators in $H^2(B_{K'})$ represent linear functionals on T which vanish on K; hence $\rho^*(RK') \subseteq P_K$ and $\mathrm{Ann}(M_{T,K}/N_{T,K}) \subseteq P_K$. Let $\mathrm{Ann}(M_{K',K}/N_{K',K}) = \cap\, q_i$ be a reduced primary decomposition in RK' with associated prime ideals in P_i. Again, since RT is flat as an RK'-module, it follows from Proposition 11 in Ch.IV, §2.6 in Bourbaki [2] that in order to prove that $\cap\, q_i \otimes_{RK'} RT$ is a reduced primary decomposition of $\mathrm{Ann}(M_{T,K}/N_{T,K})$, it is sufficient to show that all the ideals of $P_i \otimes_{RK'} RT$ are prime. Let $RK = Q[t_1,\ldots,t_\ell]$, then it is clear that $RT \cong RK'[t_1,\ldots,t_\ell]$. Here $RK'[t_1]/P_i[t_1] \simeq (RK'/P_i)[t_1]$; RK'/P_i and hence $(RK'/P_i)[t_1]$ is an integral demain, so $P_i[t_1]$ must be a prime ideal. By repetition we see that $P_i \otimes_{RK'} RT$ is prime in RT. Hence $P_i \otimes_{RK'} RT$ are the primes associated to a reduced primary decomposition of $\mathrm{Ann}(M_{T,K}/N_{T,K})$; since $P_i \subseteq RK'$ it follows that these are in P_K. Q.E.D.

PROOF OF THEOREM 4. Let $\mathrm{Ann}(M_T/N_T) = \cap\, q_i$ be a reduced primary decomposition and let $P_i = \sqrt{q_i}$. If P is a prime ideal in RT, $\mathrm{Ann}(M_T/N_T)_P \cap RT = \cap\,\{q_i;\ P_i \subseteq P\}$. Hence $P = P_i$ for one of the i's if and only if $\mathrm{Ann}(M_T/N_T)_P \cap RT \subsetneq \cap\, \mathrm{Ann}(M_T/N_T)_{P'} \cap RT$, the last intersection taken over those prime ideals P' with $P' \subsetneq P$. (Observed in Chang and Skjelbred [7]). Choose one of the P_i's and let K be the minimal subtorus with $P_K \subseteq P_i$. Let Q be any prime ideal such that the minimal subtorus L with $P_L \subseteq Q$ is equal to K. We have:

$$\mathrm{Ann}(M_T/N_T)_Q \cap RT = \mathrm{Ann}((M_T/N_T)_Q \cap RT =$$

$$\mathrm{Ann}[(M_T)_Q/(N_T)_Q] \cap RT = \mathrm{Ann}[(M_{T,K})_Q N_{T,K})_Q] \cap RT =$$

$$\mathrm{Ann}[(M_{T,K}/N_{T,K})_Q] \cap RT = \mathrm{Ann}[M_{T,K}/N_{T,K}]_Q \cap RT.$$

The first and the last equalities follow since we are dealing with finitely generated RT-modules. By the localization theorem $H_T^*(X,F)_Q \simeq H_T^*(X^K,F)_Q$; hence $(M_T)_Q \simeq (M_{T,K})_Q$ and the third equality follows. For the main step in the proof we apply Lemma 1. Since $P_K \subseteq Q$ it follows from Lemma 1 that $\mathrm{Ann}(M_{T,K}/N_{T,K})_Q \cap RT = \mathrm{Ann}(M_{T,K}/N_{T,K})$. But if $P_K \neq P_i$, this contradicts the fact that $(\mathrm{Ann}\ M_{T,K}/N_{T,K})_{P_i} \cap RT \subsetneq \cap (\mathrm{Ann}\ M_T/N_T)_{P'} \cap RT;\ P' \subsetneq P_i$. Hence $P_i = P_K$; i.e. all the associated primes are linear. The isolated primes P_i are the minimal prime ideals P containing $\mathrm{Ann}(M_T/N_T)$, i.e. they are minimal with respect to the condition that $\mathrm{Ann}(M_T/N_T)_P \cap RT \neq RT$. Again, letting K be the subtorus determined by P, we have $\mathrm{Ann}(M_T/N_T)_P \cap RT = \mathrm{Ann}(M_{T,K}/N_{T,K})$. Hence K is a maximal subtorus with respect to the condition that $M_{T,K} \neq N_{T,K}$; and this concludes the proof of Theorem 4.

REMARK. If N = (0), we get the result of Chang and Skjelbred [7] for the submodule M_T of $H_T^*(X,F)$. In this case it follows directly that $(\mathrm{Ann}\ M_{T,K})_Q \cap RT = \mathrm{Ann}\ M_{T,K}$, since it is easily shown (Theorem 1) that the map $H_T^*(X^K,F) \to H_T^*(X^K,F)_Q$ is injective. This is not sufficient to conclude that $M_{T,K}/N_{T,K} \to (M_{T,K}/N_{T,K})_Q$ is injective, and we need Lemma 1 to see that $\mathrm{Ann}(M_{T,K}/N_{T,K})_Q \cap RT = \mathrm{Ann}\ M_{T,K}/N_{T,K}$.

If X is totally non-homologous to zero in X_T, we have $H_T^*(X) \otimes_{RT} R'T \simeq H^*(F) \otimes R'T$. If M is a submodule of $H^*(F)$, Ann M_T is always a principal ideal (generated by the least common multiple of the denominators when a set of generators of M are expressed as reduced RT-rational linear combinations of elements of $H_T^*(X)$). If $(0) \neq N \subseteq M$, however, there are several examples in section 2 showing that $\mathrm{Ann}(M_T/N_T)$ is not in general a principal ideal, and the general primary decomposition is needed.

The following corollary is known (Allday and Skjelbred [1]).

PROPOSITION 1. Let X be a Poincaré duality space over Q and let T act on X. Let $F_1, \ldots, F_s$ be the connected components of $F = X^T$, let f_j be the fundamental cohomology class of F_j and 1_j the generator of $H^0(F)$. Let $M_1 = (f_j)$, $M_2 = (1_j)$. Then Ann M_1 is a principal ideal whose generator is the product of the local geometric weights at F_j with multiplicities, and the isolated prime ideals of Ann M_2 correspond to the generic isotropy subgroups of the minimal F^0-varieties connecting F_1 with other components of F.

2. GEOMETRIC WEIGHT SYSTEM FOR HIGHER COHOMOLOGY PROJECTIVE SPACES

In this section T is a torus which acts cohomology effectively on $X \sim P^n(q)$, q an even number. In the Serre spectral sequence of the fibre bundle $X_T \to B_T$ it is obvious that all differentials are zero, and X is totally non-homologous to zero. It is then easy to see that the multiplicity of a point in the variety of the defining ideal for $H_T^*(X)$ equals the Euler characteristic of the corresponding fixed point component. [18].

An immediate application of the last section then gives the following well-known result:

THEOREM 5. *Let* T *be a torus group which acts cohomology effectively on* $X \sim P^n(q)$. *Let* $e_0 \in H^n(X)$ *be a generator and* $e \in H_T^n(X)$ *a lifting of* e_0. *Then the structural equation of* $H_T^*(X) = RT[e]/\langle f(e)\rangle$ *can be written as* $f(e) = (e-w_1)^{n_1+1} \cdots (e-w_s)^{n_s+1}$, $w_i \in RT$, for $i = 1, \ldots, s$, $\sum_{i=1}^{s} (n_i+1) = n+1$.

Correspondingly $F = F_1 + \ldots + F_s$ *with* $F_i \sim P^{n_i}(q_i)$, $i=1, \ldots, s$.

The next Theorem generalizes a Theorem of Hsiang and Su [13].

THEOREM 6. *Let the rank of* T *be at least two*. *Assume* $q > 2$. *Then at most one component of* F *is of type* $P^t(p)$ *with* $2p > q$. $(t > 0)$. *Assume* $q > 4$. *Then one component of* F *is of the above type only if* F *is connected*.

PROOF. Let $j^* : H_T^*(X) \to H_T^*(F)$ be the homomorphism induced from the inclusion of F in X, and let $F = F_1 + \ldots + F_s$ with $F_i \sim P^{n_i}(q_i)$ as above. Assume that $n_1 > 0$. Then $j^*(e) = \ldots + f_1 \otimes b_1 + w_1 + \ldots + w_2 + \ldots + w_s$, ($f_1 \in H^{n_1}(F_1)$ is a generator and $b_1 \in RT$). By modifying the lifting e of e_0, if necessary, we may assume that $w_1 = 0$. Let U_i be the submodule of $H^*(F)$ generated by $H^0(F_2), \ldots, H^0(F_{i-1}), H^0(F_{i+1}), \ldots, H^0(F_s)$ and all cohomology in higher dimensions, and let $U = U_i \oplus H^0(F_i)$, $(i=2, \ldots, s)$. Then Ann $(U_T/(U_i)_T) = (w_i)$ is a principal ideal, in that case Theorem 4 implies that w_i splits as a product of linear factors. In particular $w_2 = q_2^1\omega_1^{h_1} \dots \omega_m^{h_m}$, $q_2^1 \in Q$, $\omega_j \in H^2(B_T)$, $j=1,\dots,m$. From the proof of Theorem 4 it follows that the localization $\mathrm{Ann}\ (U_T/(U_i)_{T(\omega_j)}) \cap RT = \mathrm{Ann}\ (U_{T,H_j}/(U_i)_{T,H_j}) = (\omega_j^{h_j})$, where $H_j = \omega_j^{\perp}$ is the corank one subtorus of T determined by ω_j. Hence $F(H_j) \sim P^k(2h_j) + \dots$ with $F_1 + F_2 \subseteq P^k(2h_j)$. Assume $2q_1 > q$. Since $4h_j \geq 2q_1 > q$, and $\sum_{j=1}^{m} 2h_j = q$, it follows that m=1.

Hence we may now write $w_i = q_i^1\alpha_i^h$, $i=2,\dots, s$, $q=2h$.

We may assume $n > 1$, otherwise the theorem is trivial. We claim that $s=2$ is impossible. For if $s=2$, $F(\alpha_2^1) \sim P^n(q)$, contradicting the assumption that the action os cohomology effective. Thus, if F is disconnected, $s \geq 3$, and one of the α_i, e.g. α_3 is linearly independent of α_2. If also $2q_2 > q$, then applying the same argument around F_2, we would also have $q_2^1\alpha_2^h - q_3^1\alpha_3^h = q^1\alpha^h$ for some $\alpha \in H^2(B_T)$. Let $p_i^h = q_i$ for $i = 2,3,\dots,s$ and let $\epsilon_1,\dots,\epsilon_h$ be the h-th roots of unity. Then $(p_2\epsilon_i\alpha_2)^h - (p_3\alpha_3)^h = \pi(p_2\epsilon_j\alpha_2 - p_3\alpha_3)$, $j = 1,\dots,h) = q^1\alpha^h$, by unique factorization in $RT \otimes_Q C$, $p_2\epsilon_1\alpha_2 - p_3\alpha_3$ and $p_2\epsilon_2\alpha_2 - p_3\alpha_3$ are then both complex multiples of α, which clearly contradicts linear independence of α_2 and α_3 if $h > 1$. This concludes the proof of the first statement of the theorem.

In general, if we do not assume that $2q_2 > q$, we only know that $q_2^1\alpha_2^h - q_3^1\alpha_3^h$ must split as the product of the linear factors in RT, since it generates the annihilator ideal of a quotient module as above. But this is possible only if $h = 2$ $(\epsilon_2 = -\epsilon_1)$, hence if $q > 4$ the assumption $s > 1$ leads to a contradiction. Q.E.D.

THEOREM 7. <u>Let the rank of</u> T <u>be at least four and</u> $q > 4$. <u>Then the fixed point set</u> F <u>is either connected or consists of</u> n+1 acyclic components.

PROOF. Suppose $F = F_1 + \ldots + F_s$ with $s > 1$ and $F_1 \sim P^t(d)$, $t > 0$. From the proof of Theorem 6 we have: $j^*(e) = \ldots + f_1 \otimes b_1 + 0 + \ldots + w_2 + \ldots w_s$, where each of the w_i's splits. Let V be the submodule of $H^*(F)$ generated by $H^*(F_i)$, $i \geq 2$ and $H^c(F_1)$, $c > d$; and let $W = V \oplus H^d(F_1)$. Then $\text{Ann}\,(W_T/V_T) = (b_1)$, hence b_1 also splits. Let α_1 be a weight in w_2 with multiplicity h_1, then we know that $F(\alpha_1^{\perp}) \sim P^k(2h_1) + \ldots$ for some k and $F_1 + F_2 \leq P^k(2h_1)$. By localization: $\text{Ann}\,(W_T/V_T)_{(\alpha_1)} \cap RT = \text{Ann}\,(W_{T,H_1}/V_{T,H_1}) = (\alpha_1^{k_1})$, where k_1 is the multiplicity of α_1 in b_1, hence $d + 2k_1 = 2h_1$. We can now write: $w_2 = q_2^1 \alpha_1^{h_1} \ldots \alpha_r^{h_r}$, $b_1 = q^1 \alpha_1^{k_1} \ldots \alpha_r^{k_r} \beta_r^{\ell_r} \ldots \beta_u^{\ell_u}$ with $d + 2k_i = 2h_i$, $i=1, \ldots, r$, and the α's and β's pair wise linearly independent. Here $r > 1$, since if $r = 1$: $F(\alpha_1^{\perp}) \sim P^K(q)+\ldots$ which contradicts Theorem 6 unless $F(\alpha_1^{\perp})$ is connected, i.e. $F(\alpha_1^{\perp}) \sim P^n(q)$, which again contradicts the assumption that T acts cohomology effectively. Let α_i and α_j be two weights in w_2 and let H_{ij} be the corank two subtorus annihilated by α_i and α_j. By localization: $\text{Ann}\,(U_T/U_{2T})_{(\alpha_j,\alpha_j)} \cap RT = \text{Ann}\,(U_{T,H_{ij}}/(U_i)_{T,H_{ij}}) = (\alpha_{g(1)}^{h_{g(1)}} \ldots \alpha_{g(v)}^{h_{g(v)}})$, where $\alpha_{g(1)}, \ldots, \alpha_{g(v)}$ are the α's contained in the linear span of α_i and α_j, $\text{Lin}\,(\alpha_i, \alpha_j)$. Similarly: $\text{Ann}(W_T/V_T)_{(\alpha_i,\alpha_j)} \cap RT = \text{Ann}(W_{T,H_{ij}}/V_{T,H_{ij}}) = (\alpha_{g(1)}^{k_{g(1)}} \ldots \alpha_{g(v)}^{k_{g(v)}} \beta_{p(1)}^{m_{p(1)}} \ldots \beta_{p(w)}^{m_{p(w)}})$, where those α's and β's occur which are in $\text{Lin}\,(\alpha_i, \alpha_j)$.

Now $F(H_{ij}) \sim P^k(a)$ with $a = \sum_{m=1}^{v} 2h_{g(m)} = \sum_{m=1}^{v} 2k_{g(m)} + \sum_{m=1}^{w} 2m_{p(m)} + d$. Here $2h_{g(m)} = 2k_{g(m)} + d$, and $v \geq 2$, hence $w \geq 1$. Thus we have proved that for any α_i with α_j with $i \neq j$, there is at least one

$\beta_p \in \mathrm{Lin}\ (\alpha_i, \alpha_j)$. Let K_{ij} be the corank two subtorus annihilated by α_i and β_j. By a similar argument the localized annihilator ideals have the same expressions as above, this time with the α's and β's which occur in Lin (α_i, β_j). In the above equation for a we have this time: $w \geq 1$, hence it follows that $v \geq 2$, and we have proved that for any pair (α_i, β_j), there is at least one α_k, $k \neq i$ contained in Lin (α_i, β_j).

LEMMA. <u>We can assume that the subtorus</u> K <u>annihilated by</u> $(\alpha_1, \ldots, \alpha_r)$ <u>is trivial (if necessary by renumbering</u> $F_2, \ldots, F_s$).

PROOF. We wish to prove that Lin $(\alpha_1, \ldots, \alpha_r)$ has full dimension. By localizing, it is clear that $F(K) \sim P^k(q) + \ldots$ for some k, by Theorem 6 K has rank zero for one.

Let L be the subtorus annihilated by all the weights in b_1, by a similar localization argument as above, all those weights are in Lin $(\alpha_1, \ldots, \alpha_r)$. Let H be the linear span of those weights, if H has full dimension, we are done. Otherwise H is a hyperplane. If K has rank one, it then follows that H = Lin $(\alpha_1, \ldots, \alpha_r)$. But then F(K) cannot be connected, let F_j be a component of F which is not included in the $P^k(q)$ of F(K).

Let $w_j = q_j' \gamma_1^{m_1} \ldots \gamma_b^{m_b}$, then none of the γ_i's are contained in Lin $(\alpha_i, \ldots, \alpha_r)$.

On the other hand Lin $(\alpha_1, \ldots, \alpha_r) = H \subsetneq$ Lin $(\gamma_1, \ldots, \gamma_b)$, consequently Lin $(\gamma_i, \ldots, \gamma_b)$ has full dimension. Q.E.D.

We can now conclude the proof of theorem 7 by applying Sylvester's Theorem; we use here a generalized version due to S. Hansen [9]: Let Ω be a finite set of points which spans the projective n-space. Then there is a hyperplane P, spanned by $\Omega \cap P$, and a hyperplane P' in P, spanned by $\Omega \cap P'$ such that $\Omega \cap P$ contains only one point outside P'. In our case, let Ω be given by $[\alpha_1], \ldots, [\alpha_r]$. Then, since rk $T \geq 4$, there are at least two points $[\alpha_i]$ and $[\alpha_j]$ in P', hence there is a $[\beta_k]$ in P'. Let $[\alpha_m]$ be in $(\Omega \cap P)-P'$, then there is another $[\alpha_p]$ on the line through $[\alpha_m]$ and $[\beta_k]$, in contradiction to the above theorem. Q.E.D.

REMARK. Sylvester's theorem was first applied to the cohomology theory of transformation groups by Skjelbred [16], who proved: If T acts cohomology effectively on a Poincare duality space X such that $\dim H^*(X) = \dim H^*(F)$, F has two components F^1 and F^2 such that the restriction homomorphism $H^*(X) \to H^*(F^1)$ is onto, and $\dim F^1 \neq \dim F^2$, then rk $T < 4$.

DEFINITION. The action of T on $X \sim P^n(q)$ is of spherical type if either:

1) the fixed point set F is connected, or

2) F is the disjoint union of acyclic components, and the local geometric weight systems are identical around each component

Obviously these are the actions with the simplest cohomological orbit structure. In case 1) we have $F \sim P^n(h)$, $j^*(e) = \ldots + f \otimes b$, where f is the cohomology generator of F and b splits, i.e. $b = q'w_1^{h_1} \ldots w_r^{h_r}$, $q' \in Q$, $w_i \in H^2(B_T)$. The corank one F^o-varieties are given by $F(w_i^{\perp}) \sim P^n(h + 2h_i)$. In case 2) it is clear that all F^o-varieties are connected. Hence we have $j^*(e) = q_1 n + \ldots + q_{n+1} b$, with $q_i \in Q$, $i = 1, \ldots, n+1$ and $b = q'w_1^{h_1} \ldots w_r^{h_r}$, $q' \in Q$, $w_i \in H^2(B_T)$, $2(h_1 + \ldots + h_r) = q$, $F(w_i^{\perp}) \sim P^n(2h_i)$.

DEFINITION. For an action of spherical type the geometric weight system is the above cohomology class b.

The geometric weight system completely determines the whole cohomological orbit structure of the action, in fact the following theorem is completely analogous to the case of a cohomology sphere (Hsiang [12]).

THEOREM 8. _Let_ $b = q' w_1^{h_1} \cdots w_r^{h_r}$ _be the geometric weight of a torus action on_ $X \sim P^n(q)$ _of spherical type with_ $F \sim P^n(h)$, (_if_ $h = 0$, F _is the disjoint union of_ $n+1$ _acyclic components_). _Then a subtorus_ H _is the connected component of an isotropy if and only if H is of the form_ $(w_{i(1)}, \ldots, w_{i(k)})^{\perp}$, _and then_ $F(H) = P^n(h + 2h_{i(1)} + \cdots + 2h_{i(k)} + 2h_{i(k+1)} + \cdots + 2h_{i(k+m)})$, _where_ $w_{i(1)}, \ldots, w_{i(k+m)}$ _are the weights which are in_ $\mathrm{Lin}(w_{i(1)}, \ldots, w_{i(k)})$.

Our main result for torus actions on cohomology projective spaces now follows quickly from Theorem 7.

COROLLARY 1. _Let_ T _be a torus group of rank at least 5 which act cohomology effectively on_ $X \sim P^n(q)$, $q > 4$. _Then the action is of spherical type_.

PROOF. Suppose F is not connected. By Theorem 7, $F = F_1 + \cdots + F_{n+1}$ with each F_i acyclic. Let w_i be a local geometric weight around F_1. Then $H_i = w_i^{\perp}$ has rank at least four. Since $F(H_i)$ has at least one component which is not acyclic, it follows again by Theorem 7 that $F(H_i)$ is connected, hence w_i is a local weight with the same multiplicity around all the components $F_1, \ldots, F_{n+1}$.

3. ORBIT STRUCTURE FOR ACTIONS OF COMPACT, CONNECTED LIE GROUPS ON COHOMOLOGY PROJECTIVE SPACES.

By constructing James' reduced products of spheres equivariantly, we obtain examples of actions of spherical type $X \sim P^n(q)$ with q even and n arbitrary. Let A be a Hausdorff space with base point a, and let A_n be the set of sequences in $A - \{a\}$ with no more than n terms. There is a projection p_n from A^n to A_n which excludes those terms of the n-tuple which equal a; and we give A_n the identification topology from the product A^n. If a group G acts on A with a as a fixed point, then A_n is a G-space in an obvious way.

Recall that $H^*(A_\infty)$ is isomorphic to the cohomology of the loop space of the suspension of A (James [14], [15]). If $A = S^q$, q even, we have $A_n \sim P^n(q)$. Thus, if G is a torus which acts linearly on S^q, the equivariant reduced product construction (with one of the fixed points chosen as base point) gives an action on $(S^q)_n \sim P^n(q)$ of spherical type. Obviously, this is of case 2) if and only if the given action on S^q has two isolated fixed points. The geometric weight system defined in the last section, coincides with the usual geometric weight system defined by Hsiang [12] for the original action on S^q. Thus it follows from Corollary 1 that for actions of tori of rank at least five on higher cohomology projective spaces, there is perfect correspondence between theory and examples.

It is also clear that once we have defined the geometric weight system as above, the more detailed study of the cohomological orbit structure of such actions of compact, connected Lie groups of rank at least five on $X \sim P^n(q)$ can be reduced to the corresponding problem for cohomology spheres. A systematic study of this case has been carried out by Hsiang [12]. The main idea is to restrict the action to the maximal torus and use Weyl invariance of the geometric weight system. We only sketch a few of the results which are obtained when those methods are applied to our case.

Let T be the maximal torus of a compact, connected Lie group of rank at least five, and assume that G acts cohomology effectively on $X \sim P^n(q)$, $q > 4$. We have the following straightforward observations: RT is a W-module and $RG = RT^W$. The Weyl group W acts on the fixed point set F(T). If $x \in F(T)$ and $w \in W$, the system of local geometric weights around wx are obtained from those around x by conjugation with w. On the other hand, if $b = q'w_1^{h_1} \dots w_r^{h_r}$ is the geometric weight system, those systems are both given by $\{(w_1;h_1), \dots, (w_r;h_r)\}$. Hence this is Weyl group invariant.

THEOREM 9. _Let_ $G = SU(k)$, $k > 5$, _act cohomology effectively on_ $X \sim P^n(q)$, _with_ $k(k-1) > q > 4$. _Then all orbits are finitely covered by complex Stiefel manifolds_.

PROOF. We must prove that the connected component G_x^o of any isotropy group is a standardly embedded SU(k-p) for some p. Let $b_1, \dots, b_k$ with $b_1 + \dots + b_k = 0$ be the usual coordinates on T, then W is the symmetric group on $(b_1, \dots, b_k)$. Let $b = n_1 b_1 + \dots + n_k b_k$ with n_i integers. It is easy to see that the shortest W-orbit of b occurs if $w = b_i$ for some i and the second shortest if $w = b_i + b_j$, the length of the latter (modulo sign) is $k(k-1)/2$. Hence, by the dimension restrictions of the Theorem, it is clear that only the shortest orbit can occur in the geometric weight system, which must then coincide with the weight system for a direct sum of copies of the standard representation of SU(k) and a trivial representation; i.e. the geometric weights are given by $B = \{(b_1;p),\ldots,(b_k;p)\}$ for some p. When choosing a suitable point x on an arbitrary orbit of X, one may assume that the maximal torus T_1 of G_x^o is included in T; i.e. there exist weights $b_{i(1)},\ldots,b_{i(p)}$ such that $T_1 = T_x^o = (b_{i(1)},\ldots,b_{(p)})^{\perp}$, one may as well assume that $T_x^o = (b_1,\ldots,b_p)^{\perp}$. Let D(G) be the weight system of the adjoint representation of G and $D(G)|T_1$ the restriction of this to T_1. The action of G_x along the orbit G/G_x has weight system $D(G)|T_1 - D(G_x^o)$; hence $D(G_x^o) \supset D(G)|T_1 - B|T_1$. Direct substitution then gives $D(SU(k-p)) \subset D(G_x^o) \subset D(G)|T_1$. A Lie algebra computation then gives $G_x^o \sim SU(k-p)$.

Q.E.D.

REMARK. There are analogous theorems for other classical groups; e.g.: If $G = SO(k)$, $k \geq 13$ and $k(k-3)/2 > q > 4$, then all orbits are finitely covered by real Stiefel manifolds. If $G = SP(k)$, $k \geq 6$ and $k(k-1)/2 > q > 4$, all orbits are finitely covered by quaternionic Stiefel manifolds. For more details and classification of principal orbit types, we refer to Hsiang [12]. This also contains a discussion of the important case when the geometric weight system is modelled on the adjoint action, and the Weyl group acts as a group generated by topological reflections on F(T).

REFERENCES

1. C. Allday and T. Skjelbred, The Borel formula and the topological splitting principle for torus action on a Poincaré duality space. Ann. of Math. 100 (1974) 322-326.
2. N. Bourbaki, Elements of Mathematics. Commutative Algebra. Hermann, Paris. 1972.
3. A. Borel, et al. Seminar on Transformation Groups. Ann. of Math. Studies 46. Princeton University Press 1960.
4. G. Bredon, Introduction to Compact Transformation Groups. New York-London. Academic Press 1972.
5. T. Chang and M. Comenetz, Group Actions on Cohomology Projective Planes and Products of Spheres. Preprint.
6. T. Chang and T. Skjelbred, Group Actions on Poincaré Duality Spaces. Bull. Amer. Math. Soc. 78 (1972) 1024-1026.
7. T. Chang and T. Skjelbred, The topological Schur lemma and related results. Ann. of Math. 100 (1974) 307-321.
8. T. Chang and T. Skjelbred, Lie Group actions on a Cayley projective plane and a note on orientable homogeneous spaces of prime Euler characteristic. To appear in Amer. J. of Math.
9. S. Hansen, A generalization of a theorem of Sylvester on the Lines Determined by a finite point set. Mathematica Scandinavica 16 (1965), 175-180.
10. W. Y. Hsiang, On characteristic classes and the topological Schur lemma from the topological transformation groups viewpoint. Proc. Symposia in Pure Math. XVII 105-115 (1971

11. W. Y. Hsiang, On some fundamental theorems in cohomology theory of topological transformation groups. Taita J. Math. 2, (1970), 61-87.

12. W. Y. Hsiang, On the splitting principle and the geometric weight system of topological transformation groups I. Proc. 2nd Conf. on Compact Transf. Groups. Amherst. Mass. 1971. Lecture Notes in Math. 298, 334-402. Berlin-Heidelberg-New York. Springer 1972.

13. W. Y. Hsiang and J. C. Su, On the geometric weight system of topological actions on cohomology quaternionic projective spaces. Invent. Math. 28 (1975), 107-127.

14. I. M. James, Reduced Product Spaces. Ann. of Math. 62 (1975), 170-197.

15. I. M. James, The Suspension Triad of a Sphere. Ann. of Math. 63, (1956), 407-429.

16. T. Skjelbred, Torus actions on manifolds and affine dependence relations. Preprint. IHES, Bures-sur-Yvette. 1975.

17. P. Tomter, On the geometric weight system for transformation groups on cohomology product of spheres. Prliminary report. Universität Bonn. 1972.

18. P. Tomter, Transformation groups on cohomology product of spheres. Invent. Math. 23 (1974), 79-88.

UNIVERSITY OF OSLO,

OSLO 3, NORWAY.

ON THE EXISTENCE OF GROUP ACTIONS ON CERTAIN MANIFOLDS

STEVEN H. WEINTRAUB

In this note, we show how a relatively straight-forward application of surgery theory yields existence and non-existence theorems for certain kinds of group action. The most extreme results in either direction may be summarized as follows:

THEOREM. Let M be a simply-connected manifold whose homology is torsion-free and concentrated in even dimensions. Then

a) For any odd p, if M admits a semi-free Z/p action with isolated fixed points, so does any PL-manifold homotopy equivalent to M

b) If M admits a semi-free locally smooth S^1 action, and M′, a manifold homotopy equivalent to M, also admits one with the same fixed-point data, then there are only finitely many possibilities for the PL homeomorphism type of M′. ("Fixed-point data" is defined in Definition 1.3).

This argument was developed for highly-connected M in [8]. Among the other manifolds to which it applies are products of spheres, and, more interestingly, to projective spaces (complex and quaternionic). For these latter cases, in addition to the fact that they admit interesting actions, there is a wealth of information known about their fixed-point data (see [2, Chapter VII]). In these cases we may apply an idea of Hsiang [3] to conclude

THEOREM. *Let* $M = CP^{2n}$ or HP^n, $n > p-1$, *and let* Z/p *act smoothly on* M *with* $\dim(M^{Z/p}) \leq 4n - 2p$. *Then infinitely many manifolds, homotopy equivalent to* M, *admit such an action with the same fixed-point data.*

1. GENERAL RESULTS

Let G be a cyclic group of odd order. Throughout this section M^{2n} will denote a simply-connected manifold whose homology is concentrated in even dimensions and has all of its torsion prime to the order of G. M will be the total space of a semi-free G-action with fixed-point set F. If $N(F)$ is an equivariant tubular neighbourhood of F, let $T = \overline{M - N(F)}$ and let X be the orbit space of T. We also suppose that T is simply-connected (which will be the case when $\operatorname{codim}(F) > 2$). Let $\pi : T \to X$ be the projection.

Recall from [5] that in this situation we have the following commutative diagram of exact sequences:

$$\begin{array}{ccccccc}
L_{2n-1}(G, \bigcup_{i=1}^{k} G) & \to & hT(X) & \xrightarrow{\eta_X} & [X, G/PL] & \to & L_{2n}(G, \bigcup_{i=1}^{k} G) \\
\downarrow & & \downarrow{\scriptstyle \pi^*} & & \downarrow{\scriptstyle \pi^*} & & \downarrow \\
L_{2n-1}(\{1\}, \bigcup_{i=1}^{k} \{1\}) & \to & hT(T) & \xrightarrow{\eta_T} & [T, G/PL] & \to & L_{2n}(\{1\}, \bigcup_{i=1}^{k} \{1\}) \\
 & & \uparrow{\scriptstyle r} & & & & \\
 & & hT(M) & & & &
\end{array} \qquad (I)$$

(As usual, the "G" in "G/PL" has nothing to do with the group G that is acting.)

Here the union is over the boundary components of T. $hT(T)$ denotes the homotopy triangulations of T where the boundary is allowed to vary. The map from $hT(M)$ to $hT(T)$ is that induced from deleting $\mathrm{Int}(N(F))$.

Also, note that $L_*(G, \bigcup_{i=1}^{k} G) \approx \bigoplus_{i=1}^{k-1} L_{*-1}(G)$, and $L_{2k-1}(G) = 0$ if G has odd order (by [1]), and similarly for $\{1\}$. Also, $L_*(G) \to L_*(\{1\})$ is onto.

Now, if $M' \in hT(M)$ admits a G-action with fixed point set F' such that $M' - F'$ is homotopy-equivalent to $M - F$, then $M' - N(F') = \pi^*((M' - N(F')/G))$ in $hT(T)$, i.e., $M' \in \mathrm{Im}(r^{-1}\pi^* : hT(X) \to hT(M))$. Thus, our first goal will be to determine $\mathrm{Im}(r^{-1}\pi^*)$.

LEMMA 1.1. $[(X, \partial X), G/PL] \otimes Q \to [(T, \partial T), G/PL] \otimes Q$ _is an isomorphism. Also, if_ F _is a union of isolated points,_ $[X, G/PL] \to [T, G/PL]$ _is onto while_ $[(T, \partial T), G/PL]/\mathrm{Im}([(X, \partial X), G/PL]) = G^{\chi(M)-1}$.

PROOF. Since $(X, \partial X)$ is rationally equivalent to $(T, \partial T)$ (the latter being a finite cover of the former) the first part is immediate. Similarly, $[X, G/PL]_{(q)} = [T, G/PL]_{(q)}$ for any q prime to the order of G.

Thus to prove the second statement, we need only consider the situation at a prime q dividing the order of G. Now $[X, G/PL]_{(q)} = KO^0(X)_{(q)}$, $[T, G/PL]_{(q)} = KO^0(T)_{(q)}$, and KO may be computed from the Atiyah-Hirzebruch spectral sequence, which collapses for both for dimensional reasons (since except in the top dimension their cohomology is concentrated in even dimensions). Hence, it suffices to show that $H^{even}(X : Z_{(q)}) \to H^{even}(T : Z_{(q)})$ is onto.

To compute $H^{even}(X : Z_{(q)})$, we use the spectral sequence of the cover. Now, as T is homotopy equivalent to M - a finite set of points (in fact, $\chi(M) = \sum_{i \leq 2n} \dim H_i(M)$ points, by Smith theory), $H^*(T) = H^*(M)$ for $* \leq 2n-2$, and $H^{2n-1}(T) = Z^{\chi(M)}$. In particular, $E_2^{p,q} = 0$ if either p or q is odd, except for $q = 2n-1$, so $E_2^{0,q}$ survives to E_∞ for $q < 2n-1$, so $H^*(X) \to H^*(T)$ is onto for $* < 2n-1$. Then, by considering the cohomology ladder of the pairs $(X, \partial X)$ and $(T, \partial T)$, it is easy to see that $H^*(T, \partial T)/\mathrm{Im}(H^*(X, \partial X)) = G^{\chi(M)-1}$ and the last statement follows.

THEOREM 1.2. _Let_ M _be as above, and suppose_ F _is a union of isolated points. Then, any PL-manifold homotopy equivalent to_ M _also admits a semi-free G-action with isolated fixed-point set._

PROOF. By 1.1, $\pi^* : [X, G/PL] \to [T, G/PL]$ is onto. By chasing (I), it follows that $\pi^* : hT(X) \to hT(T)$ is onto. But $r : hT(M) \to hT(T)$ is an isomorphism, since T is obtained from M by deleting the interiors of disks, while M is obtained from T by coning off the boundary spheres (a well-defined process in the PL category) and these constructions are inverses of each other.

DEFINITION 1.3. Let G act semi-freely on M. _The fixed-point data_ of the action consists of the equivariant PL homeomorphism class of N(F) and the homotopy type of the pair $(T, \partial T)$.

THEOREM 1.4. _Suppose_ $H^{4i}(M : Q) \neq 0$ _for some_ i _with_ $0 < 4i < \dim M$, _and_ M _admits a G-action with given fixed-point data. Then infinitely many manifolds of the homotopy type of_ M _do not admit semi-free G-actions with the given fixed-point data._

PROOF. The condition on $H^*(M)$ insures that there are

infinitely many manifolds of the homotopy type of M.

Suppose first that F is a union of isolated points.

To investigate actions with the same fixed-point data as the given action on M, we must fix ∂X (and hence ∂T) and so must consider the analogue of diagram (I) where X (T) is replaced by (X, ∂X) (and (T, ∂T), respectively) and the Wall groups are the absolute Wall groups. Then $\pi^* : [(X, \partial X), G/PL] \to [(T, \partial T), G/PL]$ is not onto, so $\pi^* : hT(X) \to hT(T)$ cannot be either, so neither can $r^{-1}\pi^*$.

If F is more complicated, $Im(r^{-1}\pi^*)$ can only get smaller as r^{-1} may fail to be defined.

COROLLARY 1.5. _Suppose M, as above, admits a semi-free_ S^1 _action. Then, if_ M′ _homotopy equivalent to_ M _admits one with the same fixed-point data, there are only finitely many possibilities for the homotopy type of_ M′.

If, in addition, $H^{4i+2}(M : Z/2) = 0$ _for all_ i _with_ $4i + 2 < \dim(M)$, _and_ $H^{4i}(M)$ _has no 2-torsion,_ M′ _must be_ PL _homeomorphic to_ M.

PROOF. S^1 contains every cyclic group, so the image of $[(X, \partial X), G/PL]$ in $[(T, \partial T), G/PL]$ may be made arbitrarily large by choosing $X = M/G$ for G sufficiently large. In fact, not only is the cokernel arbitrarily large but it must be isomorphic to $G^{\chi(M)-1}$ so no element of infinite order can be in all the cokernels. Hence, any element in all the cokernels must be a 2-torsion element. This proves the first claim, and the second follows since in that case there are no 2-torsion elements.

2. PROJECTIVE SPACES

Among the manifolds to which 1.5 applies are complex and quaternionic projective spaces. In this section, we show that, as opposed to the situation of circle actions, infinitely many smooth manifolds of the homotopy type of CP^{2n} or HP^n, $n > 2$, admit Z/p actions. If $M = CP^{2n}$ or HP^n, recall that Hsiang showed in [3] that there is a subgroup of finite index in $\widetilde{KO}(M)$ such that any ξ in this subgroup satisfies the conditions:

ξ is fibre-homotopically trivial

$$p_i(\xi) = 0 \quad \text{for} \quad 0 < i \le [n/2],$$

p_i denoting rational Pontrjagin class

$$L_n(\xi) + L_{n-1}(\xi)\cdot L_1(\tau(M)) + \ldots + L_1(\xi)\cdot L_{n-1}(\tau(M)) = 0,$$

$\tau(M)$ denoting the tangent bundle of M

and that every such ξ gives rise to a distinct smooth manifold of the homotopy type of M. This subgroup contains a subgroup of finite index S which is its intersection with $KO(M, M^{2n})$. S is a free abelian group of rank 2n.

THEOREM 2.1. _Suppose_ Z/p _acts on_ $M = CP^{2n}$ or HP^n, $n > p-1$, _semi-freely and smoothly and_ $\dim(M^{Z/p}) \le 4n - 2p$. _Then infinitely many manifolds of the homotopy type of_ M _admit a_ Z/p _action with the same fixed-point data._

PROOF. Let $m = \max(2n, 4n - 2p)$. Then $\bar{S} = S \cap KO(M, M^m)$ is a free abelian group of rank greater than $(p-1)/2$. Also, $KO(T, \partial T) \to \bar{S}$ is onto, and $KO(X, \partial X) \otimes Q \to KO(T, \partial T) \otimes Q$ is an isomorphism since T is a finite cover, so $KO(X, \partial X)$ is onto a subgroup of finite index $\bar{\bar{S}}$ of $\bar{S}$. Hence $[(X, \partial X)/(X, \partial X)^m, G/O]$ contains an abelian group of rank greater than $(p-1)/2 = \text{rank}(L_{4n-1}(Z/p))$, and as it is easy to see the surgery obstruction map is a homomorphism on this subgroup of $[(X, \partial X), G/O]$ it must have infinite kernel. Then this kernel gives rise to infinitely many homotopy smoothings of $(X, \partial X)$ and hence to actions on infinitely manifolds of the homotopy type of M.

REMARK. Our proof in this section could have been rephrased to look more like section 1. For finding a fibre homotopically trivial bundle is the same as finding a map of M into BO which is homotopically trivial when mapped into BG, so pulls back to a map of M into G/O. Then our conditions ensure that the surgery obstructions on X and M vanish so we obtain homotopy smoothings of M covering those of X.

3. CONCLUDING REMARKS

Similar questions to the ones we consider here have been investigated by other people. Petrie has conjectured [4] that if a homotopy CP^n admits a smooth S^1-action the homotopy equivalence must preserve the $\hat{\mathfrak{a}}$ characteristic class. Wang [6] proved that if the fixed point set of an S^1 action on a homotopy CP^n, $n > 3$, is homotopy equivalent to $CP^k \cup CP^\ell$, the underlying manifold must be, in fact, tangentially homotopy equivalent to CP^n, and announced that there are infinitely many such distinct Z/p actions (on distinct homotopy CP^n's).

Also, Petrie constructed a very interesting S^1 action (in [4]) on CP^3 where the representations at the fixed-points do not agree with those of any linear action. In light of the condition in 2.1, we ask if there can arise actions on homotopy CP^n's with fixed-point data not agreeing with those of any action on the standard CP^n. (If not, that would show many homotopy CP^n's admit no smooth Z/p actions.)

Finally, in [7], we considered in detail the question of the existence of smooth Z/p actions on homology HP^2's.

We conclude with an interesting number-theoretic question. Consider the case of actions with isolated fixed points, so the fixed-point data reduces to the normal representations at the fixed-points. For actions on spheres, with two fixed points, Atiyah-Bott showed the normal representations at the two fixed points must agree (except for sign), i.e., that they must be the normal representations of a linear action. On the other hand, Petrie's example on CP^3 with four fixed points, is distinguished from any linear action by the fact that the normal representations at the fixed points differ from those of any linear action. In [7], we found a collection of representations that could algebraically arise from an action on HP^2, with three fixed points, for the prime 11, which are distinct from those of any linear action, but such do not exist for smaller primes. Thus, at the moment, we have an anomaly, and we ask in general what is the situation for the case of three fixed points.

References

1. A. Bak, The Computation of Surgery Groups of Odd Torsion Groups. Bull. Amer. Math. Soc. 80 (1974), 1113-1116.

2. G. Bredon, Introduction to Compact Transformation Groups. Academic Press, New York, 1972.

3. W. C. Hsiang, A note on free differentiable actions of S^1 and S^3 on homotopy spheres. Ann. of Math. 83 (1966), 266-272.

4. T. Petrie, Smooth S^1 actions on homotopy complex projective spaces and related topics. Bull. Amer. Math. Soc. 78 (1972), 105-153.

5. C. T. C. Wall, Surgery on Compact Manifolds, Acadmic Press, New York, 1970.

6. K. Wang, Differentiable Circle Group Actions on Homotopy Complex Projective Spaces. Math. Ann. 214 (1975), 73-80.

7. S. Weintraub, Group Actions on Homology Quaternionic Projective Planes, to appear.

8. S. Weintraub, Semi-free Z/p actions on Highly-Connected Manifolds, Math. Z. 145 (1975), 163-185.

LOUISIANA STATE UNIVERSITY

BATON ROUGE, LA 70803

U.S.A.

PART TWO

(Summaries and Surveys)

PROPER TRANSFORMATION GROUPS

HERBERT ABELS

This is a survey. No proofs will be given. Also no historical remarks. Both can be found in the references cited at the respective places. The talk has three parts: I Definition, examples, II Results, III H_c^* , problems.

I. DEFINITIONS, EXAMPLES

Our standing assumptions will be: G is a locally compact Hausdorff topological group, X a locally compact, paracompact, locally connected, connected Hausdorff space. We look at continuous actions of G on X, $G \times X \to X$ sending (g,x) to $g \cdot x$. One very lazy action is $g \cdot x = x$ for every $x \in X$ and every $g \in G$. I consider active actions: let g_i be a net in G, divergent in G, i.e. every compact subset of G contains only a finite number of points of the net. Then $g_i x$ diverges in X. We need a bit more: $g_i K$ diverges for every compact subset K of X. This is equivalent to the following property:

DEFINITION. The action of G on X is <u>proper</u> if and only if for every pair K,L of compact subsets of X the set $\{g \in G;\ gK \cap L \neq \emptyset\}$ has compact closure in G.

Equivalently: for every pair x,y of points in X there are neighbourhoods U_x, U_y of x and y respectively such that $\{g \in G;\ gU_x \cap U_y \neq \emptyset\}$ has compact closure.

EXAMPLES.

E 0) G compact. Every action is proper. One goal of the theory is to reduce the study of proper actions of a group to that of its compact subgroups (cf. Theorem 5 below and the remarks following it). Henceforth we assume that G is not compact.

E 1) Let $X \to Y$ be a covering space and let G be the group of decktransformations. G endowed with the discrete topology acts properly on X.

E 2) Let X be a metric space. The group G of isometries endowed with the compact open topology acts properly on X. Subexamples are : let X be a Riemannian manifold, G its group of isometries, or let X be a bounded domain in C^n (more generally in a Stein manifold), G its group of biholomorphisms. In each of these cases G acts properly on X, when endowed with the compact open topology.

E 3) Let G = X (this time X need not be connected nor locally connected). G acts on X by left translations. This action is proper. This example can be varied: let H be a closed subgroup of G, K a compact subgroup of G. The action of H on G/K induced by left translations is proper.

E 4) Let G be a discrete group with a finite set E of generators. Consider the Cayley diagram C(G,E), i.e. the graph whose vertices are the elements of G and whose edges are the pairs

(g,gh), $g \in G$, $h \in E \cup E^{-1}$. The geometric realization X of C(G,E) is a proper G-space under the action induced by left translation.

There are some questions arising naturally

Q 1: Given X. Is there a proper action of some (non compact) G on X? If so, X is not compact, trivially. A necessary condition on X is given by Theorem 1 below.

Q 2: Given G and X. Is there a proper action of G on X? If yes, classify them. An answer to this question is given in Theorem 5 and the discussion following it for G almost connected.

Q 3: Let G act properly on X. Are there relations between G and X? Any such gives a necessary condition for yes to Q 2.

II. RESULTS

Some of the nicest results are obtained in terms of the ends of X.

A compactification Y of X is called 0-dimensional if $Y \setminus X$ has (Menger-)dimension zero, equivalently if $Y \setminus X$ is totally disconnected.

The <u>endpoint compactification</u> (or Freudenthal compactification) X^+ of X is a 0-dimensional compactification of X such that

1) If Y is a 0-dimensional compactification of X, there is a continuous map $X^+ \to Y$ making the diagram

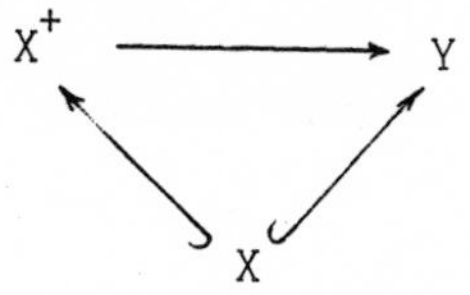

commutative.

Or equivalently:

2) $X^+ \setminus X$ does not disconnect X^+ locally, i.e. given a neighbourhood V in X^+ of a point $y \in X^+ \setminus X$ there do not exist two disjoint open subsets U_1, U_2 of $V \cap X$ such that $U_1 \cup U_2 = V \cap X$ and $y \in \bar{U}_1 \cap \bar{U}_2$. (E.g. if y has a neighbourhood base of V's such that $V \cap X$ is connected). X^+ exists and is essentially unique. We call $E(X) = X^+ \setminus X$ the <u>space of ends</u> of X and $e(X) = \#E(X)$ <u>the number of ends of X</u>.

EXAMPLES.

E 1) $R^+ = (0,1)^+ = [0,1]$ by 2). So $e(R) = 2$.

E 2) $(R^n)^+ = S^n$, $n > 1$. So $e(R^n) = 1$ for $n > 1$. R^n is a group manifold so a proper R^n-space, $n \geq 1$.

E 3) More generally: let M be a compact connected manifold without boundary of dimension > 1, let E be a finite subset of M or a convergent sequence of points plus limit point. Then for $X = M \setminus E$ we have $M = X^+$. E.g. $C \setminus (Z + iZ)$ has infinitely many ends and is a proper $Z + iZ$-space.

In the examples we have spaces X admitting proper actions of a non-compact group with $e(X) = 1, 2, \infty$.

For all our theorems the hypothesis will be:

HYPOTHESIS. Let X be a proper G-space for a non-compact group G.

THEOREM 1 [1]. $e(X) = 1, 2$ <u>or</u> ∞. <u>If</u> G <u>is</u> <u>connected</u>, $e(X) \leq 2$.

Now we turn to the question: what are the groups G in the cases $e(X) = 1, 2$ and ∞ resp? In case $e(X) = 1$ nothing can be said because: if X is a proper G-space, so is $X \times X$ and $e(X \times X) = 1$.

For $e(X) = 2$ there is

THEOREM 2 [1]. Suppose $e(X) = 2$. Then the orbit space $G\backslash X$ is compact and G contains a cocompact discrete subgroup $\cong Z$.

The structure of these groups G is known: G contains a compact normal subgroup K such that G/K is an extension of R or Z by a group of ≤ 2 elements.

If $e(X) = \infty$, we cannot expect much in general: let G be a discrete infinite group acting properly on a manifold X of dimension > 1. Then X minus the orbit of one point is a proper G-space Y with $e(Y) = \infty$, by E 3. So we have to pay special attention to the limit points: let X be a proper G-space. A point $y \in X^+ \setminus X$ is called a limit point of G, if y is in the closure of some orbit Gx, $x \in X$. Any limit point is in the closure of any orbit Gx, $x \in X$. Let $L(G)$ be the set of limit points.

One can show that $X_1 = X^+ \setminus L(G)$ is a proper G-space and $X_1^+ = X^+$. So again: $\#L(G) \leq 2$ or infinite, etc.

THEOREM 3 [2]. Suppose $\#L(G)$ is infinite. Then G is

(a) a free product $G_1 *_K G_2$ amalgamated over a compact open subgroup K of G_1 and G_2, $K \neq G_1$, $K \neq G_2$, or

(b) an HNN-extension $G = G_1 *_\alpha$, where K is compact open subgroup of G_1, α is an open continuous injective homomorphism $\alpha : K \to G_1$, and $K \neq G_1$ or $\alpha(K) \neq G_1$. The group $G = G_1 *_\alpha$ has the generators and relations $\langle G_1, x;\ x k x^{-1} = \alpha(k) \text{ for } k \in K\rangle$.

This theorem is the version for topological groups of a theorem of Stallings [11].

Under the hypotheses of Theorem 3, $L(G)$ is a minimal G-space (i.e. the closure of any G-orbit in $L(G)$ is $L(G)$) except if (b) occurs and $K = G_1$, $\alpha(K) \neq G_1$ or $K \neq G_1$, $\alpha(K) = G_1$. In this exceptional case there is a fixed point in $L(G)$ and any other orbit is dense. The exceptional case cannot occur if G is locally connected (see [4]).

I mention in passing that there are relations between ends of spaces and ends of groups ([7,8,9,10,2]).

It would be useful if one could define higher-dimensional analogues of the end-point compactification.

Returning to the examples at the beginning of this section II, we have seen that - with the notations of E3 - for $M = S^n$ there are E with $1 \leq \#E \leq 2$ or $\#E = \infty$ such that $X = M \setminus E$ admits a proper action. What about other manifolds $M = X^+$?

THEOREM 4. _If_ X^+ _is an n-manifold without boundary,_ X^+ _is homeomorphic to_ S^n.

The theorem proved in [6] actually says: if one limit point $y \in X^+$ has a neighbourhood homeomorphic to an n-ball, then $X^+ \simeq S^n$.

For the next result we need the following fact. We call a group G almost connected if the factor group G/G_o is compact, where G_o is the connected component of the neutral element.

FACT. An almost connected group G contains a maximal compact subgroup K. Any compact subgroup of G is conjugate to a subgroup of K. The space G/K is homeomorphic to a euclidean space.

The dimension of this euclidean space is unique, because any two maximal compact subgroups of G are conjugate. We call it the non-compact dimension of G : $n_c(G)$.

THEOREM 5 [3]. Let X be a proper G-space, G almost connected. Let K be a maximal compact subgroup of G. Then there is a G-map $f: X \to G/K$.

It is easy to see that this map turns X into a locally trivial fibre bundle over G/K with fibre $S = f^{-1}(K/K)$ and structure group K. Since G/K is contractible this fibre bundle is trivial. So

$$X \simeq S \times R^{n_c(G)}.$$

So a necessary condition for G acting properly on X is that $R^{n_c(G)}$ is a direct factor of X. It is easy to see that this condition is actually sufficient. More precisely: let E(G;X) be the set of equivalence classes of proper actions of G on X, two actions being equivalent if there is a G-homeomorphism between them. Then

$$E(G;\ X) = \cup\ E(K;\ S)$$

the disjoint union being taken over all spaces S such that $X \simeq S \times R^{n_c(G)}$, one such out of every class of homoeomorphic S. This formula reduces the problem of determining all proper actions of G on X to two problems: (1) determine all S such that $X \simeq S \times R^{n_c(G)}$ up to homeomorphism, (2) determine the equivalence classes of actions of a maximal compact subgroup K on S.

III. H_c^*, PROBLEMS.

By $H_c^*(\cdot;k)$ we mean sheaf theoretic cohomology with coefficients in the constant sheaf with stalks a field k and compact supports. Let us define $n_c(X;k) = \inf\{i; H_c^i(X;k) \neq 0\}$, $n_c(X;k) = \infty$ if $H_c^*(X;k) = 0$. We consider $n_c(X;k)$ as the non-compact dimension of X. This is in keeping with our earlier notation: $n_c(G) = n_c(G;k)$

for every k, if G is almost connected. More generally: the formula $X \simeq S \times \mathbb{R}^{n_c(G)}$ and the Künneth theorem imply:

COROLLARY. *Hypotheses and notations as in Theorem 5. Then*

$$n_c(X;k) \geq n_c(G)$$

and

$$n_c(X;k) = n_c(G) \quad \textit{if and only if } G\backslash X \textit{ is}$$

compact.

Note the qualitative implication: given two proper actions of such G on X then either both orbit spaces are compact or both are non-compact.

This raises the *question* whether the corresponding statement for arbitrary G and X is true. It is easily seen to be not true in general.

COUNTEREXAMPLE: inject the free group on two generators into itself $i : G \to G$ such that $i(G)$ is of infinite index in G. Then in I Example 4 the orbit space under the natural action of G on X is compact whereas the proper action $(g,x) \to i(g)x$ has non-compact orbit space.

There is the following interesting result:

THEOREM 6 (H.C. Wang [12]). *Let* X *be a differentiable manifold with finite dimensional cohomology algebra* $H^*(X;k)$ *with coefficients in a finite field* k. *Given a group* G *and two differentiable, proper and effective actions of* G *on* X. *Then either both orbit spaces are compact or both are non-compact.*

Note that the hypotheses imply that G is actually a Lie group.

The corresponding theorem for differentiable manifolds with boundary is false. A counterexample can be constructed starting with the above counterexample G,X. Embed X in R^3, take Y a regular neighbourhood of X. Extend the action of G to Y. Y is contractible.

But the conclusion of the above theorem holds if X is a differentiable manifold with boundary and $\dim_k H_c^*(X;k) < \infty$ for k a finite field (see below). This assumption is equivalent to Wang's for orientable differentiable manifolds by Poincaré duality.

CONJECTURE. Let X be a space with $\dim_k H_c^*(X;k) < \infty$, k a finite field. Given two proper and effective actions of a group G on X. Then either both orbit spaces are compact or both are non-compact.

By analysing the proof of the above theorem of Wang's one obtains the following result. Let us call an action of G on X <u>p-free</u>, p a prime number, if the isotropy subgroups G_x are all finite and their cardinality is not divided by p.

THOEREM 7. *Let* G *be a Lie group*, k *a finite field of characteristic* p. *There is an integer* $N_c(G;p) \geq 0$ *such that for any proper p-free G-space* X *with* $\dim_k H_c^*(X;k) < \infty$ *we have*

$$n_c(X;k) \geq N_c(G;k)$$

and

$$n_c(X;k) = N_c(G;k) \text{ if and only if } G\backslash X \text{ is compact.}$$

In particular the conjecture holds for these actions on such X.

A way to prove the conjecture in general for a Lie group G would be to construct a 'freeing action' over any proper G-action.

QUESTION: Given a proper G-space X. Is there a free G-action on some principal fibre bundle Y over X with fibre a compact connected Lie group?

The answer is no in general. I do not know the answer for discrete G.

The answer is yes in the following two cases: (1) G acts properly, effectively and differentiably on a differentiable manifold X (with or without boundary): endow X with an invariant Riemannian metric. If X is oriented, the bundle of oriented orthonormal frames will do. If X is not oriented, embed X into X x X diagonally. Apply the procedure to the manifold X x X (with corners) and restrict the bundle to X. (2) G is a closed subgroup of an almost connected Lie group, X arbitrary. This is a consequence of a theory of universal proper G-spaces [5].

A yes to the above question for G discrete and G\X compact would be implied by a positive answer to the

QUESTION. Given a simplicial complex S, on which a discrete G acts simplicially and with finite isotropy groups. Suppose G\X finite. Can the geometric realization of X be G-embedded into a differentiable manifold, on which G acts differentiably, effectively and properly?

Since there is not enough time to talk about relations between H_c^* and H^* for proper G-spaces nor about relations between $H_c^*(X)$ and $H^*(G;kG)$ for discrete G, I will stop here.

REFERENCES

1. H. Abels. Enden von Räumen mit eigentlichen Transformationsgruppen. Comm. Math. Helv. 47 (1972) 457-473.
2. H. Abels. Specker-Kompaktifizierungen von lokal kompakten topologischen Gruppen. Math. Z. 135 (1974) 325-361.
3. H. Abels. Parallelizability of proper actions, global K-slices and maximal compact subgroups. Math. Ann. 212 (1974) 1-19.
4. H. Abels. On a problem of Freudenthal's. Preprint.
5. H. Abels. A universal proper G-space. Preprint.
6. H. Abels and A. Lauer. A characterization of the sphere and euclidean space by transformation groups. Math. Z. 146 (1976) 137-142.
7. H. Hopf. Enden offener Räume und unendliche diskontinuierliche Gruppen. Comm. Math. Helv. 16 (1943/44) 81-100.
8. H. Freudenthal. Über die Enden diskreter Räume und Gruppen. Comm. Math. Helv. 17 (1944) 1-38.
9. E. Specker. Die erste Cohomologiegruppe von Überlagerungen und Homotopie-Eigenschaften dreidimensionaler Mannigfaltigkeiten. Comm. Math. Helv. 23 (1949) 303-333.
10. E. Specker. Endenverbände von Räumen und Gruppen. Math Ann. 122 (1950) 167-174.
11. J. Stallings. Groups of cohomological dimension one. Proc. Symposia in Pure Math. 17. Applications of Categorical Algebra (New York 1968). 124-128. Providence, R.I., AMS 1970.

12. H.C. Wang. A remark on co-compactness of transformation groups. Amer. J. Math. 15 (1973) 885-903.

UNIVERSITY OF BIELEFELD

BIELEFELD, WEST GERMANY

PROBLEMS ON GROUP ACTIONS ON Q MANIFOLDS

R. D. ANDERSON

1. PROPERTIES OF Q AND OF Q-MANIFOLDS.

We let Q denote the Hilbert cube: the countable infinite product of closed intervals. A Q-manifold is a separable metric space with an open cover of sets homeomorphic to open subsets of Q. Since Q is compact, every Q-manifold is locally compact. It is also an ANR.

Topological properties of Q and/or of Q-manifolds have been studied extensively and intensively over the past 10 or 12 years by T.A. Chapman, James E. West, Raymond Y-T Wong, L. Siebenmann, R.D. Edwards, S. Ferry, Z. Cerin, R.M. Schori, the author and several others. Introductory work on compact group actions on Q and on Q-manifolds have resulted in many interesting examples and one significant partial result by Wong. It is the purpose of this paper to survey the known results and techniques, to pose several open problems and to propose various areas for investigation.

The reader is referred to the book by Bessager and Pelcznski on Selected Topics in Infinite-Dimensional Topology (published, 1975, in Warsaw) and the CBMS Expository Lecture Notes on Q-Manifolds by T.A. Chapman (to be published in 1976 by the AMS) for background information and problems in infinite-dimensional topology.

We review below some of the properties of Q and of Q-manifolds used in posing problems and developing intuition about Q. In general, we do not give specific references but usually do attribute authorship to others by name.

PROPERTIES OF Q. O.-H. Keller proved in 1931 that every compact convex infinite-dimensional subset of ℓ_2 is homeomorphic to $(\approx)$ Q. As a corollary, cone (Q) $\approx$ Q. We define a Z-<u>set</u> in Q to be a closed set K in Q such that for any $\epsilon > 0$ there is a map of Q into $Q \setminus K$ within ϵ of the identity map of Q onto Q. For example, any compact subset of $S \subset Q$ or of $Q \setminus S$ is a Z-set in Q where $S = \Pi_{i > 0}(-1,1)_i$ and $Q = \Pi_{i > 0}[-1,1]_i$.

The so-called Homeomorphism Extension Theorem (HET) for Q asserts that any homeomorphism h of a Z-set K_1 onto a Z-set K_2 of Q can be extended to a homeomorphism H of Q onto Q. Furthermore, by Barit, for any $\epsilon > 0$, this can be done with $d(H,id) < d(h,id) + \epsilon$. As a corollary of HET, Q is homogeneous (i.e. for any points A, B $\in$ Q, there is a homeomorphism h of Q onto Q with $h(A) = B$) a result originally proved by Keller. From Q $\approx$ cone (Q) and Q being homogeneous, it follows (by considering the cone point) that for any point p of Q and any $\epsilon > 0$, there is a neighbourhood U of p of diameter $< \epsilon$ such that $Q \setminus U \approx \text{Bdry } U \approx \text{Cl } U \approx Q$. These facts help make Q and Q-manifolds easy to deal with; for example the HET refers to closures of nice neighbourhoods, their boundaries, and their complements.

In 1963-64 it was proved that any triad (letter Y) or more generally any dendron D was a Q-factor, i.e. $D \times Q \approx Q$. Later

West (1969-1970) showed that any contractible polyhedron K was a Q-factor and, generalizing earlier results of the author, that any countable infinite product of non-degenerate Q-factors were homeomorphic to Q. In 1975, R.D. Edwards characterized Q-factors by showing that every compact AR was a Q-factor (and obviously any Q-factor in an AR). As noted below, these results give us very powerful tools in identifying and studying group actions on Q.

PROPERTIES OF Q-MANIFOLDS. West (1969-1970) proved that for any countable locally finite polyhedron K, $K \times Q$ is a Q-manifold. Later Chapman proved that for any Q-manifold M, there exists a countable locally finite polyhedron K such that $M \approx K \times Q$. In effect, this is the triangulation theorem for Q-manifolds. He also showed that for any two Q-manifolds M_1 and M_2, $M_1 \approx M_2$ if and only if for $M_1 \approx K_1 \times Q$ and $M_2 \approx K_2 \times Q$, K_1 and K_2 are of the same infinite simple homotopy type. For the compact Q-manifold case, Chapman's results showed the invariance of Whitehead torsion, the first major application of Q-manifold theory to problems arising in algebraic topology. Chapman had earlier shown that if two Q-manifolds each admitted a half-open interval factor, i.e. $M_1 \approx M_1' \times [0,1)$ and $M_2 \approx M_2' \times [0,1)$ for some Q-manifolds M_1' and M_2', then $M_1 \approx M_2$ if and only if $M_1 \sim M_2$ (or $M_1' \sim M_2'$) where '$\sim$' means is 'homotopy equivalent to'. It is known that many (non-compact) Q-manifolds do not admit half-open interval factors.

Later West used Q-manifold theory to show that each compact metric ANR has the homotopy type of a finite polyhedron thus giving another application of Q-manifold theory to problems arising elsewhere in topology. Subsequently, R.D. Edwards showed that $X \times Q$

is a Q-manifold if and only if X is a locally compact metric ANR, thus characterizing Q-manifold factors and including West's results on the homotopy type of compact metric ANR's as a corollary.

As noted below, these various results on Q-factors and on Q-manifold factors provide many interesting examples and raise many interesting questions concerning group actions on Q-manifolds. Since Q-manifolds are becoming an increasingly important tool in other areas of topology, the study of further properties of Q-manifolds seems very much in order.

2. GENERATING GROUP ACTIONS ON Q

(i) Any action of a group G on a Q-factor X induces a natural action on $Q \approx X \times Q$ defined by $g(x,g) = (gx,g)$. In other words, G acts on the product with the given action on the first factor and with the identity action on the second.

(ii) Any action of a group G on a non-degenerate Q-factor X induces an infinite product action on Q. For each $i > 0$, let $X_i \approx X$ and let G_i be G considered as acting on X_i. Then G acts on $\Pi_{i>0} X_i \approx Q$ defined by $gx = (g_i x_i)$ where $g \in G$ and $x = (X_i) \in \Pi_{i>0} X_i$. Observe that if (G,X) has a single fixed point then so does the induced action (G,Q). However, if (G,X) has two fixed points, then (G,Q) has a Cantor set of fixed points.

(iii) For each i, let X_i be a non-degenerate AR and let $G = G_i$ be a group acting on X_i. Then G acts on $Q = \Pi_{i>0} X_i$ by $gx = (g_i x_i)$ as in (ii) above. Here we do not assume the X_i's are homeomorphic. If (G_1, X_1), for example, has an arc of fixed points and for $i > 1$, (G_i, X_i) has a single fixed point, then (G,Q) has an arc of fixed points.

(iv) If G is a compact Lie group, then, as a space, G is an ANR and cone (G) is a compact AR. G acts canonically on cone (G) by left translation on levels. In this case (G, cone (G)) has exactly the cone point, p, as a fixed point and cone $(G) \setminus \{p\}$ consists of a half-open interval of invariant copies of G. Combining this procedure with that of (ii) above we get any compact Lie group acting on Q semi-freely with a single fixed point; in other words every orbit other than the fixed point is a full copy of G.

(v) Any compact group which is an inverse limit of Lie groups (e.g. a Cantor or solenoidal group) can act on Q with a single fixed point. We write $G = \varprojlim G_i$. For $g = (g_i) \in \varprojlim G_i$ and for $x = (x_i) \in \Pi_{i > 0}$ cone (G_i), define (G,Q) by $gx = (g_i x_i)$ where g_i acts on x_i as in (iv) above. Under this action it is not in general true that each orbit other than that of the fixed point is full. For example any orbit in the cone(G_1) is reproduced in the product by crossing such orbit with the fixed (cone) points of the other factors. The fact that Cantor and solenoidal groups are easily seen to operate effectively on Q (and thus on any Q-manifold M regarded as K x Q with the identity action on K) makes the infinite-dimensional case dramatically different from that of finite-dimensional manifolds.

(vi) The following example is due to Chapman. Let Q′ be regarded as cone(Q) with vertex p. Then any group G action on Q induces group G acting on cone (Q) = Q′ by G acting on levels of the cone with each fixed point of Q under G generating an arc of fixed points in cone(Q). If (G,Q) is semi-free with a single fixed point, then (G,Q′) is semi-free with an arc α of fixed points. Now the arc α is a Z-set in cone Q and it is known that if such an arc α is shrunk to a point with all other points preserved, then the resultant space Q* is homeomorphic to Q′ (and to Q) and the induced action gives G acting on Q* semi-freely with a single fixed point. Each level of the cone Q above p produces an invariant copy of Q under (G,Q*). Deleting the fixed point p* we have both $Q^* \setminus p^*$ and the orbit space of $Q^* \setminus p^*$ admitting half open interval factors. As noted below in Section 4, this will let us show the equivalence of appropriate finite group actions.

In general, the process of shrinking Z-set arcs or other Z-sets of trivial shape in Q has produced many examples and theorems in infinite dimensional topology. Indeed, Edward's results referred to above as characterizing Q-factors and Q-manifold factors employ another form of the shrinking of Z-sets of trivial shape.

3. SOME EXPLICIT EXAMPLES OF Z/3 ACTIONS

We consider various Z/3 actions on Q by way of illustration. (a) From (iv) above on Lie group actions we have Z/3 acting on Q with a single fixed point with Q regarded as a product of triods (letters 'Y'). The action rotates each factor about its branch point. (b) From (ii) above we may consider Q as a product of discs with Z/3 rotating each disc about its centre. (c) From (ii) above we may consider Q as a product of 6-ods with Z/3 rotating each 6-od about its branch point. Each full orbit hits just 3 branches. (d) For any ANR, X, which admits a free Z/3 action we can cone X to get an AR and a semi-free Z/3 action with a single fixed point, the cone point, as in (iv) above. Any countable infinite product of such AR's is Q and by (iii) we can induce a semi-free Z/3 action on Q with a single fixed point. (e) For any of the actions a) to d) we can cone the action as in (vi) above, shrink the arc of fixed points to a point and get a different appearing Z/3 action on Q with a half-open interval of invariant copies of Q intersecting only in their common fixed point.

It is known that all these actions are, in fact, equivalent (up to conjugation) as we point out in the next section.

4. THE BASIC ARGUMENT.

Here we show the equivalence of certain semi-free finite group actions on Q with a single fixed point.

Let $Q_o = Q \setminus \{p\}$ where p is the fixed point of the action. Then we have G acting freely on Q_o, a contractible non-compact Q-manifold homeomorphic to $Q \times [0,1)$. Elementary covering space apparatus is applicable as noted. Let G act freely on $Q_o(1)$ and $Q_o(2)$, assumed to be copies of Q_o. We have the diagram

$$\begin{array}{ccc} Q_o(1) & \text{———} & Q_o(2) \\ \downarrow & & \downarrow \\ Q_o(1)/G & & Q_o(1)/G \end{array}$$

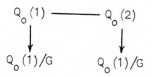

with the two orbit spaces (called reduced orbit spaces) being non-compact Q-manifolds. They are also Eilenberg-MacLane spaces and since their homotopy groups are identical, they are homotopy equivalent. When these (and other) conditions are known to insure that $Q_o(1)/G$ and $Q_o(2)/G$ are homeomorphic then any homeomorphism can be lifted to a homeomorphism of $Q_o(1)$ and $Q_o(2)$ and with the one point compactifications of $Q_o(1)$ and $Q_o(2)$ and of $Q_o(1)/G$ and $Q_o(2)/G$ we have the induced actions on $Q(1)$ and $Q(2)$ equivalent.

In general, two non-compact Q-manifolds of the same homotopy type need not be homeomorphic; we need them to be of the same infinite simple homotopy type (Chapman). However, there are known results applicable to the cases cited in sections (2) and (3) which give us partial results.

I. By Chapman's theorem which says that two Q-manifolds which are of the same homotopy type and which also admit half-open interval factors are homeomorphic, we know that any two semi-free finite group actions of G on Q with a single fixed point are equivalent if the reduced orbit spaces admit half-open interval factors. Thus for a given finite group G, any two of the coned and shrunken arc actions described in Section 2(vi) (or discussed in Section 3 e) are necessarily equivalent. Indeed, the coning of Q and shrinking of the resultant fixed arc produces new actions of G which are necessarily equivalent regardless of the actions we start with. If there is an exotic action, it is tamed by the coning and shrinking process.

The partial result that handles all the known cases cited above is due to Wong - who originally stated it only for special cases such as involutions. Wong defines a semi-free action with single fixed point as <u>trivial at the fixed point</u> if there exist arbitrarily small invariant contractible neighbourhoods of the fixed point. Then he proves that any two semi-free actions of a finite group G on Q with single fixed point and trivial at that point are equivalent. By selecting small invariant neighbourhoods of the fixed point in finitely many factors and crossing with the other factors we get Wong's condition of triviality at the fixed point in all known examples of the types discussed above.

5. OPEN PROBLEMS CONCERNING GROUP ACTIONS

The following represent only a few of many questions inherent in the preceding discussion.

a) The open problem which has attracted the most attention is the problem as to whether every two semi-free actions of a finite group G on Q with a single fixed point are equivalent. Note that Q has the fixed point property. Wong, West, Hastings and David Edwards have all worked on this problem using various techniques including proper homotopy equivalences and/or some pro-homotopy apparatus. The problem appears to be very delicate. Final results are unknown even for the Z/2 (involution)case.

b) An associated problem is whether every semi-free action of a finite group G on Q with a single fixed point is trivial at the fixed point (in the sense of Wong as in Section 4).

c) Problems related to those of a) or b) include the more general questions of equivalence of two semi-free actions of a finite group G with an arc of fixed points or with some other elementary set of fixed points. Such questions potentially introduce different techniques since the one-point compactification of spaces and orbit spaces under free actions are no longer automatically available.

d) The inverse limit procedures of section 2(v) produce group actions on Q which are not semi-free. A natural question is to ask whether Cantor or solenoidal group actions on Q can exist with the added provisions of semi-free actions with a single fixed point. In other words, does there exist a free Cantor group (or solenoidal group) action on Q_0?

e) A major area of investigation is the determination of conditions under which (finite) group actions on Q-manifolds admit factorizations into an action on a countable locally finite polyhedron K and an action on Q. For example, does every involution on $S^2 \times Q$ admit a factorization into an involution on S^2 by an involution on Q? In this connection, it should be recalled that Q-manifold factorizations as $K \times Q$ are not unique. For example, $S^1 \times Q$ is homeomorphic to the Möbius band $\times$ Q. Thus the proper framework for questions should be the existence of a factorization into a finite-dimensional factor by Q rather than the use of a particular factorization. Also it seems likely that actions may be describable in terms of maps on or of factors rather than of actions on factors per se. To date, these questions of actions on Q-manifolds have had no serious inquiry. It seems likely that an interesting and useful theory can be found.

LOUISIANA STATE UNIVERSITY, BATON ROUGE, LA 70803, U.S.A.

A NON-ABELIAN VIEW OF ABELIAN VARIETIES

L. AUSLANDER, B. KOLB AND R. TOLIMIERI

INTRODUCTION

The role of nilpotent groups, particularly the Heisenberg group, in the theory of Theta functions and Abelian varieties has been implicitly known for some time and made explicit by Weil and Mumford. In this paper we will describe a way of naturally attaching a rational nilpotent Lie group N(T) and an automorphism J of N(T) to a complex torus T. In terms of the morphism properties of the pair (N(T),J) and certain homogeneous spaces of N(T), we will be able to:

1. Give a necessary and sufficient condition for T to be an abelian variety.

2. If T is an abelian variety, fully describe the field of meromorphic functions of the variety.

In order to make the presentation as striking as possible we will develop the theory without mentioning meromorphic functions, line bundles or theta functions. At the end of the paper we will relate the objects used by us to the classical ones.

1. THE RATIONAL NILPOTENT LIE ALGEBRA OF CERTAIN MANIFOLDS.

Since our first construction is of some independent interest we will present it in a general setting and later see how to specialize it to the cases of particular interest to us.

Let M be a manifold. Let $H^*(M,R)$ be its real cohomology algebra and let $H^+(M)$ be the algebra of elements of positive degree. Clearly $H^+(M)$ is a nil-associative algebra as the product of m elements, $m > \dim M$, is zero. Now given any real associative algebra A we can convert the underlying vector space A into a Lie algebra L(A) by defining the Lie bracket of x, $y \in L(A)$ by the formula

$$[x,y] = xy - yx$$

where the multiplication on the right hand side is in the algebra A. We will denote the Lie algebra associated to $H^+(M)$ by L(M). Since L(M) is nilpotent the exponential mapping makes L(M) into a Lie group which we will denote by N(M). One can verify easily that L(M) is a 2-step nilpotent Lie algebra whose centre contains all elements of even order in $H^+(M)$.

We will be interested in this construction in precisely two special cases: first, when M is a compact Riemann surface of genus $g > 0$, in which case L(M) can be identified with the Heisenberg algebra L of dimension (2g+1). Explicitly L has a basis $X_1,\dots,X_g$, $Y_1,\dots,Y_g,Z$ such that the Lie bracket satisfy

$$[X_i,Y_i] = Z \quad i = 1,\dots,g$$

$$[X_i,X_j] = [Y_i,Y_j] = 0 \quad 1 \le i,\ j \le g$$

$$[X_i,Y_j] = 0 \quad i \neq j .$$

Clearly Z spans the centre z of L and L/z is abelian. Further, and this will be very important in our later considerations, the complex structure on M is classically shown, using holomorphic forms, to induce an automorphism J of L. The automorphism J restricted to the centre is the identity mapping and J induces an automorphism on L/z whose square is minus the identity. Second, when $T = V/D$, where V is a 2n dimensional real vector space and D is a lattice subgroup so that T is a 2n dimensional torus. In this case the cohomology algebra can be identified with the Grassman algebra $\wedge(V^*)$, where V^* is the vector dual of V.

Let us now return to our general setting. If M has torsion free cohomology, $H^+(M,Z) \subset H^+(M,R)$ gives a basis of $H^+(M,R)$ which is a Z submodule. Thus $H^+(M,Z)$ determines a rational form of the Lie algebra L(M). Hence the Lie group N(M) has discrete co-compact subgroups Γ and so we may naturally assign the compact homogeneous spaces or, so called, nilmanifolds $N(M)/\Gamma$ to the pair $H^+(M,Z) \subset H^+(M,R)$.

Let us return again to the torus $T = V/D$. Let $v_1,\ldots,v_{2n}$ be a basis of $H_1(T,Z)$. We can use this basis to determine a rational structure on $H^1(T,R)$. In our algebraic model for $H^1(T,R)$, this reflects itself in an identification of $\wedge^+(V^*)$ and $\wedge^+(V)$ and an identification of L(T) and $L(\wedge^+(V))$. Clearly the elements of $\wedge^+(D) \subset \wedge^+(V)$ generate a lattice subgroup and so determine a rational form of $\wedge^+(V)$ which we will call the D-rational form.

Now assume that V has a complex structure J, i.e., an automorphism of V such that $J^2 = -I$, where I is the identity mapping. Then J induces an automorphism of $\wedge^+(V)$ and hence of L(T). We will denote this automorphism of L(T) also by J.

The following definition is suggested by the classical facts about Jacobi varieties of compact Riemann surfaces.

DEFINITION. Let L be the 2n+1 dimensional Heisenberg algebra and let L(T) be defined as above. A Lie algebra morphism $\varphi : L(T) \to L$ is said to be of type H with respect D if

(1) φ is a D-rational epimorphism (i.e., the kernel K of is a D-rational subspace).

(2) $\varphi \mid \wedge^1(V)$ is 1 - 1

(3) $\varphi \mid \wedge^j(V) = 0$ for $j > 2$

(4) If K equals the kernel of φ then K is J invariant

(5) The J induced map

$$\bar{J}: \wedge^2(V)/K \cap \wedge^2(V) \to \wedge^2(V)/K \cap \wedge^2(V)$$

is the identity.

Since the underlying vector space of L(T) and $\wedge^+(V)$ coincide, we will use this identification in discussing L(T). Let z denote the centre of L. Because $\wedge^2(V)$ is central in L(T) and φ is an epimorphism, we have easily that $\varphi(\wedge^2(V)) = z$. Once we fix an isomorphism of z to the reals, we can identify $\varphi|\wedge^2(V)$ with a linear functional, i.e., an element of $\wedge^2(V^*)$ we identified with the cohomology group $H^2(T,R)$.

Let us now indicate some of the immediate consequences for $T = V/D$ of the existence of an H morphism $\varphi : L(T) \to T$. Let N denote the group obtained from L by means of the exponential mapping. We will call N the Heisenberg group. We have, by identifying an abelian Lie group with its Lie algebra, the following compositions of group homomorphism

$$N(T) \to N \xrightarrow{p} N/z = V$$

where we will use z to denote the centre of N also. Consider $D \subset V$

and let $v_1,\ldots,v_{2n}$ be a basis of D. Let $x_1,\ldots,x_{2n}$ be elements of N such that

$$p(x_i) = v_i \quad i = 1,\ldots,2n \quad .$$

Let Δ_D be the subgroup of N generated by the elements $x_1,\ldots,x_{2n}$. Then because of our assumption about the rationality of the mapping φ it follows that Δ_D is a discrete subgroup of N whose quotient space is compact.

DEFINITION. We call Δ_D a cover of D, and use $\{\Delta_D\}$ to denote the set of covers of D.

We close this section with the easy, but crucial observation that

$$\begin{array}{c} N/\Delta_D \\ \downarrow \\ V/D \end{array}$$

is a principal circle bundle under the action of the group $z/\Delta_G \cap z$.

2. SOME ANALYTIC FACTS ABOUT N/Δ_G

In this section N will denote as usual a Heisenberg group, z the centre of N, and Γ a discrete subgroup of N such that

a) $\Gamma\backslash N$ is compact.

b) The commutator subgroup of Γ, $[\Gamma,\Gamma]$, equals $\Gamma \cap z$.

Clearly condition b is satisfied by covering groups Δ_D.

Consider the compact homogeneous space of right co-sets of Γ in N, $\Gamma\backslash N$. Then $\Gamma\backslash N$ admits a unique probability measure μ invariant under the action of N. We get a unitary representation R of N on $\mathfrak{L}^2(\Gamma\backslash N)$, where we have fixed the measure μ, by setting

$$R_v F(\Gamma w) = F(\Gamma wv) , \quad F \in \mathcal{L}^2(\Gamma\backslash N)$$

$v, w \in N$. It is clear that we have a representation of the Lie algebra L on $C^\infty(\Gamma\backslash N)$ and, indeed, if L_C denotes the complexification of L then we have a representation of L_C on $C^\infty(\Gamma\backslash N)$. We also consider functions F on $\Gamma\backslash N$ as functions on N which are fixed under left translations by Γ. In general, if F is a function in N and $v \in N$, we set

$$(L_v F)(w) = F(v^{-1}w) \quad w \in N .$$

Thus if $L_\gamma F = F$, $\gamma \in \Gamma$, we may view F as a function on $\Gamma\backslash N$ and we will call such functions Γ periodic functions. We may restrict R to $z(N)/z(\Gamma) = T^1$, the circle group, and so we get a unitary representation, $R|T^1$, of the circle group T^1. It is then well known that

$$\mathcal{L}^2(\Gamma\backslash N) = \oplus\Sigma H_m(\Gamma)$$

where the sum is the orthogonal direct sum and where

$$H_m(\Gamma) = \{F \in \mathcal{L}^2(\Gamma\backslash N) \mid F(\gamma\xi) = \exp\left(\frac{2\pi i m}{\beta(\Gamma)}\gamma\right)F(\xi)\}$$

when $\xi \in N$, $\gamma \in z$ and $\beta(\Gamma)$ is the positive generator of $z \cap \Gamma$.

REMARK. It is important to note that if $F, G \in H_m(\Gamma)$ then F/G is, at least formally, a function on $N/\Gamma z = T$ where Γz is the subgroup generated by Γ and z in N.

Now if we have an automorphism J of L such that

$$J^2 = -I \bmod z$$

J determines a direct sum decomposition

$$L_C = V_i(J) \oplus V_{-i}(J) \oplus z_C$$

where $V_{\pm i}(J)$ is the eigenvalue $\pm i$ subspaces of L_C with respect to

J and z_C is the centre of L_C.

Let $A(J) = \{F \in C^\infty(N) | V_i(J)F = 0\}$. In general $V_i(J)$ determines a set of n linearly independent differential operators on N. We will be interested in $F \in A(N)$ which are also Γ periodic. To fix notation let

$$\Theta(J,\Gamma) = A(J) \cap \mathcal{L}^2(\Gamma \backslash N)$$

$$\Theta_m(J,\Gamma) = A(J) \cap H_m(\Gamma).$$

It may happen that $\Theta(J,\Gamma)$ is trivial.

3. STATEMENT OF MAIN RESULTS

Let $T = V/D$ be a complex torus. Then T is called an abelian variety if there exist enough meromorphic functions on T to separate points when T is an abelian variety. We will use $\mathcal{F}(T)$ to denote the field of meromorphic functions on T.

THEOREM 1. $V/D = T$ _is an abelian variety if and only if there exists an_ H _morphism_ $\varphi : L(T) \to N$ _with respect to_ D _such that if_ Δ_D _is a cover of_ D _then_ $A(J,\Delta_D)$ _is not empty._

THEOREM 2. _Let_ T _be an abelian variety and let_ Δ_D _be a cover of_ D _in_ N. _Let_ F, $G \in \Theta_m(J,\Delta_D)$ _and let_ $\mathcal{F}$ _be the field generated by_ $\{F/G | F, G \in \Theta_m(J,\Delta_D), m \in Z\}$ _then_ $\mathcal{F} = \mathcal{F}(T)$.

We will close this announcement with an interpretation of the set $\{\Delta_D\}$.

Let $A \in \Lambda^2(V^*)$ which we identified with $H^2(V/D,R)$. A will be called the Chern class of some line bundle over V/D if and only if $A \in H^2(V/D,Z)$, i.e. A takes integral values on $D \wedge D$. Clearly $A \in H^2(V/D,Z)$ determines N and therefore $\{\Delta_D\}$ and vice versa.

THEOREM 3. *There is a natural* 1 - 1 *correspondence between* $\{\Delta_D\}$ *and isomorphism classes of holomorphic line bundles over* V/D *with Chern class* A.

Let Δ_1 and Δ_2 be elements of $\{\Delta_D\}$. Then there exists an inner automorphism of N taking Δ_1 to Δ_2. The group of inner automorphisms of N can be identified with V^*. This gives us a natural representation of V^* as a transitive group acting on $\{\Delta_D\}$. Let $D^* \subset V^*$ be such that $D^*(\Delta_D) = \Delta_D$.

THEOREM 4. *We can identify* V^*/D^* *with* $\{\Delta_D\}$. *If* V/D *is an abelian variety,* V^*/D^* *is an abelian variety, the so called dual variety of* V/D.

C.U.N.Y

NEW YORK, N.Y. 10031

U.S.A.

NON COMPACT LIE GROUPS OF TRANSFORMATION AND INVARIANT OPERATOR MEASURES ON HOMOGENEOUS SPACES IN HILBERT SPACE *

M. P. HEBLE

INTRODUCTION

Our approach is somewhat as follows. By suitable assumptions about the space and the group acting on it, we try to ensure the existence of some (if not all) of the ingredients which occur when we assume compactness. This, admittedly, is a slight departure from the programme indicated by Smale (cf. [10], [11], [6]). We are able to arrive at some results on integrability of Pfaffian systems in Hilbert and Banach spaces and on invariant operator-measures, and these results as well as their proofs, are completely non-trivial, owing to the technical difficulties involved. The detailed proof are contained in [3], and our results are analogous to some earlier results due to S. S. Chern ([2], [8], [9]).

I. Our basic space is H^n, the n-fold direct sum of H with itself ($n \geq 1$) where H is an infinite-dimensional real Hilbert space. The group G of transformations acting on H^n is an

* This paper is a short summary (without proofs) of the author's paper [3] which will be published in "Advances in Mathematics" (Academic Press Inc.)

r-parameter Lie group of diffeomorphisms $H^n \xrightarrow{\text{onto}} H^n$ where the parameter r-tuple $a = (a_1, \ldots, a_r)$ is a point in the "parameter space" $B^r = \underbrace{B \oplus \ldots \oplus B}_{r}$ $(r \geq 1)$, B being some infinite-dimensional Banach space, and G is subject to the assumptions (A) - (D) below. In B^r let B_i be the subspace consisting of vectors $(a_1, \ldots, a_r)$ with $a_j = 0$ for $j \neq i$, so that $B^r = B_1 \oplus \ldots \oplus B_r$. For $1 \leq h < r$ we shall denote the subspace $B_{h+1} \oplus \ldots \oplus B_r$ of B^r by ${}_{r-h}B$.

(A) Each element $g \in G$ is a diffeomorphism of $H^n \xrightarrow{\text{onto}} H^n$, satisfying: (i) $f(g; x) = g(x) = x' \in H^n$ for $x \in H^n$, $g \in G$; (ii) $g_1, g_2 \in G$, $x \in H^n \Rightarrow g_1[g_2(x)] = (g_1g_2)x \in H^n$; (iii) $x' = g(x) = f(g; x)$ is simultaneously continuous with respect to both $x \in H^n$ and $g \in G$; (iv) the "coordinates" $x' = (x_1', \ldots, x_n')$ of the transformed point are functions $x_i' = \varphi_i(x_1, \ldots, x_n; a_1, \ldots, a_r)$, $i = 1, \ldots, n$, which are C^1-smooth (Fréchet) with respect to the x_i's and a_i's, and C^2-smooth with respect to the a_i's for each fixed $x \in H^n$.

This assumption implies that each $a \in B^r$ yields an element $T_a \in G$ and T_a maps a given $\alpha \in B^r$ into $\alpha' \in B^r$ by: $T_{\alpha'} = T_a T_\alpha$. The group of all these transformations $T_a : B^r \to B^r$, $a \in B^r$, is called the <u>parameter group</u>, denoted $\mathbf{P}$.

(B) The transformation in G corresponding to $a = (0, \ldots, 0)$ is assumed to be the identity: $\varphi_i(x_1, \ldots, x_n; 0, \ldots, 0) = x_i$, $i = 1, \ldots, n$. Next we assume that if $a = (a_1, \ldots, a_r)$ is changed by the addition of a vector, denoted $da = (da_1, \ldots, da_r)$, then the diffeomorphisms T_{a+da}, T_a are related by: $T_a^{-1} T_{a+da} = T_\epsilon$ where $\epsilon = \epsilon(a; da) = [\epsilon_1, \ldots, \epsilon_r] = [\sum_{k=1}^{r} A_k^1(a)da_k, \ldots, \sum_{k=1}^{r} A_k^r(a)da_k]$, with $A_k^i \in \mathcal{B}(B_k, B_i)$, $1 \le i, k \le r$. The operator matrix $((A_k^i))$ represents an element of $\mathcal{B}(B^r)$. Our assumption implies that, for the parameter point $a + da$, we have "in the first approximation", or in other words (" $\doteq$ " means "approximately equal"):

$$x_i' \doteq x_i + \sum_{j=1}^{r} D_{n+j}\, \varphi_j \Big|_{a=0} \epsilon_j, \quad \text{where} \quad D_{n+j}\, \varphi_i \Big|_{a=0}$$

is the first partial derivative of φ_i with respect to a_j at $a = 0$, and the difference between the left side and the right side is $o(\|da\|)$ for each $i = 1,\ldots,n$. We say that the ϵ_i define the infinitesimal transformation T_ϵ.

(C) By "frame" or "coordinate system" $\mathfrak{R}$ we shall mean either the ordered n-tuple of Hilbert spaces $H_1, H_2,\ldots,H_n$ with $H^n = H_1 \oplus \ldots \oplus H_n$, or the image of this under a transformation: $x \to \Theta x + \alpha$, Θ being an orthogonal transformation: $H^n \xrightarrow{\text{onto}} H^n$, and α a fixed point independent of x.

(D) A transformation $T_a \in G$ transforms a certain "absolute" or "initial" frame $\mathfrak{R}_o$ into a frame $\mathfrak{R}_a$: $\mathfrak{R}_a = T_a \mathfrak{R}_o$. By this we shall mean, more explicitly, the following. We consider a complete orthonormal system $\{\varphi_\alpha^{(i)}\}_{\alpha \in I}$ in each H_i in the coordinate system $\{H_1,\ldots,H_n\}$. The "absolute" coordinate system $\mathfrak{R}_o$ will then consist of the orthogonal subspaces $H_1^0, \ldots, H_n^0$ which are images of $H_1, \ldots, H_n$ under an orthogonal transformation $\Theta^0 : H^n \xrightarrow{\text{onto}} H^n$ carrying the c.o.s. $\{\varphi_\alpha^{(i)}\}_{\alpha \in I}$ into the c.o.s. $\{\varphi_\alpha^{(i,0)}\}_{\alpha \in I}$, $i = 1,\ldots,n$. Similarly the frame $\mathfrak{R}_a$ will consist of the orthogonal subspaces $H_1^a,\ldots,H_n^a$ which are images, respectively, of $H_1,\ldots,H_n$ under an orthogonal transformation $\Theta^a : H^n \xrightarrow{\text{onto}} H^n$, carrying the c.o.s. $\{\varphi_\alpha^{(i)}\}_{\alpha\in I_n}$ into $\{\varphi_\alpha^{(i,a)}\}_{\alpha\in I}$.

Let $x \in H^n$, and let $x' = T_a x$. We represent the coordinates of x, x' with respect to $\mathfrak{R}_0$ by x_i^0, $x_i'^0$ respectively ($i = 1,\dots,n$), and the coordinates of the same points x, x' with respect to $\mathfrak{R}_a$ by x_i^a, $x_i'^a$, respectively. Using $\{\varphi_\alpha^{(i,0)}\}_{\alpha\in I}$, $\{\varphi_\alpha^{(i,a)}\}_{\alpha\in I}$, the meaning of the assumption (D) can be stated as a lemma.

LEMMA 1. $x_i'^a = x_i^0$, $i = 1, \dots, n$.

Next, the transformation carrying $\mathfrak{R}_a$ into $\mathfrak{R}_{a+da}$ is $T_{a+da}\, T_a^{-1}$ which, when referred to $\mathfrak{R}_a$ becomes $T_a^{-1}(T_{a+da}\, T_a^{-1})T_a = T_a^{-1}\, T_{a+da} = T_\omega$ say, where we use the symbol $\omega = [\omega_1, \dots, \omega_r]$. Since we make extensive use of exterior differential forms in the paper, it is convenient at this point to recall the basic concepts. For any Banach spaces E, F, and any integer $k \geq 1$, we shall denote by $\mathcal{B}_k(E; F)$ the B-space of continuous k-linear mappings $E \to F$ $[\mathcal{B}_1(E, F) = \mathcal{B}(E; F)]$, and by $\mathcal{A}_k(E; F)$ the closed linear subspace in $\mathcal{B}_k(E; F)$ consisting of continuous k-linear alternating mappings $E \to F$ [for $k = 1$, $\mathcal{A}_1(E; F) = \mathcal{B}(E; F)$]. For a given set $U \in E$, and given integer $p \geq 0$ or $p = \infty$, we shall denote by $\Omega_k^{(p)}(U; F)$ the vector space of <u>differential forms of degree</u> k, <u>defined in</u> U <u>with values in</u> F (or briefly, <u>differential</u> k-<u>forms defined in</u> U with values in F), <u>of class</u> p, i.e. mappings

$$\omega : U \to \mathcal{A}_k(E; F)$$

which are of class C^p. If $U = E$, we shall simply talk of <u>differential</u> k-<u>forms on</u> E <u>into</u> F, <u>of class</u> C^p. We shall only consider the case $E = F = B^h$, for a given integer $h : 1 \leq h \leq r$. Differential k-forms of class C^p on B^h will be called C^p-<u>smooth</u>

differential k-forms on B^h, and for $k=1$ such forms will be called C^p-smooth Pfaffian forms on B^h. (For standard results and techniques concerning exterior differential forms, one may refer to Cartan [1], or Lang [5].)

Returning to the group G, the Pfaffian forms $\omega_i(a; da)$ are called the relative components of G (or of $\mathfrak{R}_a$) or the forms of Maurer-Cartan. With the help of the transformation T_ω corresponding to $\omega = [\omega_1, \ldots, \omega_r]$ we can relate the value of φ_i for a small change δx in x to the value of φ_i for a small change da in a. In fact, we have

LEMMA 2. In the first approximation $\varphi_i(x + \delta x; a) = \varphi_i(x; a + da)$, $i = 1, \ldots, n$.

Other useful properties of the ω_i's are stated in the next lemmas.

LEMMA 3. The ω_i's are "independent" meaning that the operator-matrix $((A_k^i))_{1 \le i,k \le r}$ in the above representation of the $\omega_i(a; da)$ has the following properties:

(i) $[[A_k^i(a)]]$ yields a bijective linear mapping: $B^r \xrightarrow{\text{onto}} B^r$, for each fixed $a \in B^r$;

(ii) the operator equation

$$[A_1(a), \ldots, A_r(a)] = [C_1(a), \ldots, C_r(a)] \cdot [[A_k^i(a)]]$$

for a given r-tuplet $[A_1(a), \ldots, A_r(a)]$ with $A_j(a) \in \mathfrak{B}(B_j, B^r)$, has a unique solution $[C_1(a), \ldots, C_r(a)]$ where $C_j(a) \in \mathfrak{B}(B_j, B^r)$.

LEMMA 4. The ω_i are invariant under $\mathbf{P}$. Any Pfaffian form invariant under $\mathbf{P}$ is a linear combination with constant operator-coefficients of the ω_i.

In view of Lemma 4, we also use the name "forms of Maurer-Cartan" for any linear combinations $\Omega_i = \sum_{k=1}^{r} C_k^i \omega_k$, $i = 1, \ldots, r$ such that $[[C_k^i]]$ yields a bijective linear mapping $B^r \xrightarrow{\text{onto}} B^r$.

The proofs of the major results of this section (Theorem 1, and Corollary 1.1 to be stated below) require a clarification of the concept of a "smooth (r-h)-variety in B^r". This is done via the next few definitions.

Let Q be the set of all C^2-smooth diffeomorphisms $S : B^r \xrightarrow{\text{onto}} B^r$ with the properties:

$$\text{for } \alpha \in B^r, \quad S(\alpha + d\alpha) - S(\alpha) = [[A_k^i(\alpha)]]d\alpha + O[\|d\alpha\|],$$

where (i) for each $\alpha \in B^r$, $[[A_k^i(\alpha)]]_{1 \le i,k \le r}$ yields a linear homeomorphism $B^r \xrightarrow{\text{onto}} B^r$, (ii) $[[A_k^i(\alpha)]]$ is C^1-smooth with respect to $\alpha \in B^r$ and (iii) for each $\alpha \in B^r$, $[[A_k^i(\alpha)]]$ satisfies the contention of Lemma 3. Then we define a (r-h)-variety in B^r as follows (cf. [4]; [12], pp. 304-307).

DEFINITION 1. By a <u>smooth</u> (r-h)-<u>variety</u> in B^r we shall mean the image of ${}_{r-h}B+w$ where $w \in B^r$, under a diffeomorphism $S : B^r \xrightarrow{\text{onto}} B^r$, where $S \in Q$.

We also need to define appropriate "coordinates" or "parameter r-tuples" for points in B^r with reference to the varieties V_i^h, W_j^{r-h} (defined below). Also, we need to express a given (r-h)-variety by means of convenient vector equations. We proceed as follows.

Let $S \in Q$, $S(B^h) = V^h$, $S({}_{r-h}B) = W^{r-h}$. Then both V^h, W^{r-h} are submanifolds in B^r. Since S is a diffeomorphism $B^r \xrightarrow{\text{onto}} B^r$, the tangent space at O, $T(B^r)_O$ i.e. B^r, is linearly homeomorphic under the differential $dS(0)$ to the tangent space of $S(B^r)$ $(= B^r)$ at $a = S(0)$, and $dS(0)(B^h)$ is linearly homeomorphic to $T(V^h)_a$. Hence, further $T(V^h)_a$, $T(W^{r-h})_a$, the tangent spaces of V^h, W^{r-h} at $a = S(0)$ are direct summands in B^r (cf. [7], p. 22, Theorem and Corollary). A further induction argument shows that B^r is diffeomorphic to the product manifold $V_1^h \times \ldots \times V_h^h \times W_1^{r-h} \times \ldots \times W_{r-h}^{r-h}$, where $V_i^h = S(B_i)$, $i = 1,\ldots,h$, $W_j^{r-h} = S(B_{h+j})$, $j = 1,\ldots,r-h$. We shall denote a point in this product manifold by $(y_1,\ldots,y_h;\ y_{h+1},\ldots,y_r)$ with $y_i \in V_i^h$, $1 \le i \le h$, $y_{h+j} \in W_j^{r-h}$, $1 \le j \le r-h$.

DEFINITION 2. The r-tuples $(y_1,\ldots,y_r)$ corresponding to points $y \in B^r$, where $y_i \in V_i^h$, $1 \le i \le h$, $y_{h+j} \in W_j^{r-h}$, $1 \le j \le r-h$, will be called "coordinates" or "parameter r-tuples" determined by the varieties V_i^h and W_j^{r-h}.

Next, let $S \in Q$, and let P_i be the projection of B^r onto B_i, $1 \le i \le h$. Then the (r-h)-variety $V = S({}_{r-h}B + w)$ will be given by vector equations:

$$P_i \cdot S^{-1}(y) - \xi_i \equiv P_i \cdot S^{-1}[(y_1,\ldots,y_h;\ y_{h+1},\ldots,y_r)] = 0,\ 1 \le i \le h,$$

where $\xi_i = P_i(w)$, $1 \le i \le h$, and y is a variable point on V. We can write these vector equations as:

$$F_i(y_1,\ldots,y_r;\ \xi_1,\ldots,\xi_h) = 0, 1 \le i \le h.$$

These equations then have the following property $\mathbb{B}$:

$\mathbb{B}$: the functions F_i are C^2-smooth jointly with respect to $y = (y_1,\ldots,y_r)$ and $\xi = (\xi_1,\ldots,\xi_h)$ varying over $B^r = V_1^h \times \ldots V_h^h \times W_1^{r-h} \times \ldots \times W_{r-h}^{r-h}$ and over $B^h = B_1 \oplus \ldots \oplus B_h$ respectively;

the operator-matrix $[[d_j F_i]]_{1\le i,j\le h}$ yields a linear homeomorphism: $H^h \xrightarrow{\text{onto}} B^h$ where $H^h = T(V^h)_a$; and the operator-matrix $[[d_{r+j} F_i]]_{1\le i,j\le h}$ yields a linear homeomorphism $B^h \xrightarrow{\text{onto}} B^h$.

Next let $\mathcal{F}$ be a family of smooth (r-h)-varieties with the property $\mathbb{A}$:

$\mathbb{A}$: through each point of the space B^r there passes one and only one variety of $\mathcal{F}$.

Then clearly $\mathcal{F}$ can be parametrised by points $\xi = (\xi_1,\ldots,\xi_h) \in B^h$, and the functions F_i in the vector equations $F_i = 0$ $(1 \le i \le h)$ which represent the varieties of $\mathcal{F}$, are C^2-smooth jointly with respect to $y \in B^r$ and $\xi \in B^h$ and further have the property $\mathbb{B}$ above.

Thus we agree upon the

CONVENTION. A family $\mathcal{F}$ of smooth (r-h)-varieties V satisfying $\mathbb{A}$ will be denoted by h vector-equations.

$$F_i(y;\ \xi) = 0, \quad 1 \le i \le h,$$

where: (i) ξ varying over B^h is taken to parametrise the family $\mathcal{F}$; (ii) for each fixed $\xi \in B^h$, $y \in B^r$ varies over the particular variety V corresponding to the parameter-point ξ, and the F_i have the property $\mathbb{B}$ above.

DEFINITION. The system of vector differential euqations

$$\Omega_1 = 0,\ldots,\Omega_h = 0,$$

where $\{\Omega_1,\ldots,\Omega_h\}$ form a system of Pfaffian forms, is called <u>completely integrable</u> if and only if there exists a family $\mathcal{F}$ of smooth (r-h)-varieties $V = V_{r-h}$ satisfying $\mathbb{A}$ above and such that the equations $\Omega_i = 0$, $1 \le i \le h$ are identically satisfied when the point moves on a variety $V \in \mathfrak{F}$. And under these circumstances we call the varieties $V \in \mathfrak{F}$ the integral varieties of the system $\Omega_1 = 0, \ldots, \Omega_h = 0$.

These troublesome details having been satisfactorily taken care of, we can now state the two major results of this section.

THEOREM 1. _If the Pfaffian system_ $\Omega_1(a;\, da) = 0, \ldots, \Omega_h(a;\, da) = 0$ _of vector equations where the_ Ω_i _are_ h _independent Pfaffian forms, is completely integrable and is invariant under_ $\mathbb{P}$, _then it is equivalent to the system found by equating to zero (vector),_ h _linear combinations of the_ $\omega_i(a;\, da)$'s.

COROLLARY 1.1. _A necessary condition for the Pfaffian (vector) system_ $\Omega_1 = 0, \ldots, \Omega_h = 0$ _to be completely integrable, is that the exterior derivatives vanish:_ $\Omega_i' = 0$.

II. Let G be an r-parameter Lie group of transformations on H^n as in I. We now consider a collection of submanifolds π in H^n, and assume that G transforms the elements π transitively. Consider a fixed element π_0 and assume that the subgroup g which leaves π_0 fixed is a closed Lie subgroup depending upon a set of elements belonging to a (r-h)-variety in B^r. We denote the subgroup as well as the (r-h)-variety related to it, by the same symbol. Then the transformations $T_s \in G$ or $s \in G$, for short, such that $s \notin g$, will map $g = g_{r-h}$ into sg_{r-h} so, that these sg_{r-h} fill up all of the parameter space B^r, any two being either distinct or identical. By virtue of the results of I, the varieties sg_{r-h} are shown to be h-parameter integral varieties of the system formed by the first h forms of Maurer-Cartan of G viz. of the system $\omega_1 = 0, \ldots, \omega_h = 0$.

We next consider a suitable field Σ (which might in special cases by a sigma-field) of subsets of these h-parameter varieties, each of which can be specified by a particular parameter h-tuple $\xi = (\xi_1, \ldots, \xi_h)$, so that the problem of finding a measure on sets of elements π invariant under G is equivalent to that of finding a measure on sets of points ξ, invariant under $\mathbb{P}$. The differential form $\Omega_h(z) = \omega_1(z) \wedge \ldots \wedge \omega_h(z)$, defined for $z \in \Sigma$, is invariant under $\mathbb{P}$ but may not always define a density of measure on Σ. The major theorem of this section is:

THEOREM 2. _A necessary and sufficient condition for_ Ω_h _to be a density of measure for the elements_ π _is that its exterior derivative vanish:_ $d\Omega_h = 0$.

III. _In this section we assume_ $n > 1$. We consider the _orthogonal group_ $O(n)$ of orthogonal transformations T acting on $H^n \xrightarrow{\text{onto}} H^n$, and first show that a convenient parametrisation of $O(n)$ is by points $a = (a_1, \ldots, a_r) \in B_r$, where $B = \mathcal{B}(H)$, for $r = n(n-1)/2$. We then consider the Grassmanian G_n^h, i.e. the collection of h-subspaces of H^h, $1 \leq h < n$, where by h-subspaces in H^n we mean an image of $H^h = H_1 \oplus \ldots \oplus H_h$ under an orthogonal transformation $T \in O(n)$. A suitable sigma-field Σ in G_n^h is formed by considering a fixed h-subspace viz. $V_0 = H^h = H_1 \oplus \ldots \oplus H_h$, and the h-subspaces V' which are images of V_0 by elements of $O(n)$ are collected into subsets

$$B_{t,t'} = \{V' \text{ a h-subspace in } H^n \,|\, t \leq \hat{d}(V_0, V') < t'\},$$

$0 \leq t,\ t' \leq 1$, where $\hat{d}$ is the metric:

$$\hat{d} = k \cdot \tilde{d},$$

$$\tilde{d}(V_0, V') = \max\{d(V_0, V'),\ d(V', V_0)\},$$

$$d(V_0, V') = \sup_{\substack{u \in V_0 \\ \|u\|=1}} \inf_{v \in V'} \|u-v\|,$$

and k is a positive constant so chosen that

$$\sup\{k \cdot \tilde{d}(V_0, V') \mid V' \text{ a h-subspace in } H^n\} = 1.$$

We take Σ to be the sigma-field formed from such sets $B_{t,t'}$. By the results of I, II we prove:

THEOREM 3. $\mathbb{T}_{t'} - \mathbb{T}_t$ _is an element of measure on_ $B_{t,t'}$, _where_

$$\mathbb{T}_t = \{T_t \in O(n) \mid T_t(V_0) = V' \text{ and } \hat{d}(V_0, V') = t\}.$$

This collection $\{\mathbb{T}_t\}$ is shown to have the semi-group property: $\mathbb{T}_s \cdot \mathbb{T}_t = \mathbb{T}_{s+t}$, $s, t \geq 0$, the semi-group $\{\mathbb{T}_t\}_{t \geq 0}$ is uniformly equi-continuous: $\mathbb{T}_t = e^{tC}$, and the operator-jet measure thus determined on Σ is countably additive.

REFERENCES

1. H. Cartan, _Differential forms_. Boston: Houghton-Mifflin, 1970.

2. S. S. Chern, On integral geometry in Klein spaces. _Ann. of Math_. 43 (1942), 178-189.

3. H. P. Heble, Integral-geometric measures on homogeneous spaces in Hilbert space. To appear in _Advances in Mathematics_.

4. N. H. Kuiper, Les Varietes Hilbertiennes. Seminaire de Mathematique Superieure, Montreal, 1969.

5. S. Lang, Differentiable manifolds. New York: Addison-Wesley, 1972.

6. Z. Nitecki, Differentiable dynamics. M.I.T. Press, Cambridge, Mass., London, England, 1971.

7. R. S. Palais, Lectures on the topology of infinite-dimensional manifolds. Brandeis University, 1964-65.

8. L. A. Santalo, Introduction to integral geometry. Actualites Sci. Ind. No. 1198, Paris, Hermann, 1953.

9. L. A. Santalo, Integral geometry, pp. 147-193 in "Studies in Global Geometry and Analysis", ed. S. S. Chern, MAA Studies in Maths., Vol. 4.

10. S. Smale, Differentiable dynamical systems. Bull. Amer. Math. Soc. 73, 747-817, 1967.

11. R. F. Williams, Non-compact Lie group actions. Proc. Conf. Transf. Groups, New Orleans 1967, pp. 441-445, Springer, Berlin and New York, 1968.

12. H. Whitney, Complex Analytic Varieties. Addison-Wesley, 1972.

UNIVERSITY OF TORONTO,
TORONTO, ONTARIO,
M5S 1A1 CANADA.

APPROXIMATION OF SIMPLICIAL G-MAPS BY EQUIVARIANTLY NON DEGENERATE MAPS

SOREN ILLMAN *

Let G be a finite group. A simplicial G-complex consists of a simplicial complex X together with a G-action $\varphi: G \times X \to X$ such that the map $g: X \to X$ is a simplicial homeomorphism for every $g \in G$. We say that a simplicial G-complex X is an <u>equivariant simplicial complex</u> if the following conditions are satisfied.

1. For every subgroup H of G we have that if $s = \langle v_o, \ldots, v_n \rangle$ is a simplex of X and $s' = \langle h_o v_o, \ldots, h_n v_n \rangle$, where $h_i \in H$, $i = o, \ldots, n$, also is a simplex of X then there exists $h \in H$ such that $hv_i = h_i v_i$, $i = 0, \ldots, n$.

2. For any simplex s of X the vertices $v_o, \ldots, v_n$ of s can be ordered in such a way that we have $G_{v_n} \subset \ldots \subset G_{v_o}$.

Here G_x denotes the isotropy subgroup of G at x. We call G_{v_n} the principal isotropy subgroup of the simplex s and G_{v_o} the maximal isotropy subgroup of s. The above conditions are purely technical in the sense that any simplicial G-complex can be made into an equivariant simplicial complex by passing to barycentric subdivisions. (A simplicial G-complex satisfying condition 1 is

* Partially supported by an NSF grant at the Institute for Advanced Study, Princeton, 1974-75.

called regular by Bredon [1; p.116].) Let Z be an equivariant subcomplex of the equivariant simplicial complex X. Consider the following property.

PROPERTY P. Let s be a simplex of X and let v be some vertex of s such that $v \in Z$. Then if v' is any vertex of s such that $G_v \subsetneq G_{v'}$, we also have $v' \in Z$.

We say that Z is strongly full in X if Z is full in X and satisfies Property P. This condition is also purely technical in the same sense as above.

We shall now define the notion of an equivariant combinatorial manifold. Let $\rho: G \to O(n)$ be an orthogonal representation. By $R^n(\rho)$ we denote euclidean space R^n together with G-action through ρ. Furthermore we define

$$D^n(\rho) = \text{convex hull of } \{\pm g\, e_i \mid g \in G,\ i = 1,\dots,n\},$$

$$S^{n-1}(\rho) = \partial\, D^n(\rho) .$$

(Here $e_1, e_2, \dots, e_n$ denote the standard unit vectors in R^n.) The G-spaces $S^{n-1}(\rho)$ and $D^n(\rho)$ can be triangulated such that they become equivariant simplicial complexes. An equivariant simplicial complex M is an equivariant combinatorial manifold if the following holds. For every vertex v of M there exists an orthogonal representation $\tau: G_v \to O(n)$ and a G_v-equivariant p.ℓ. homeomorphism

$$\alpha: Lk(v, M) \to S^{n-1}(\tau) .$$

We say that a simplicial G-map $f: X \to M$ is equivariantly nondegenerate if f embeds equivariant simplexes. (An equivariant simplex of X is a G-subset of the form Gs, where s is an ordinary simplex of X.) It is easy to see that a simplicial G-map $f:X \to M$ which is isovariant (i.e., $G_{f(x)} = G_x$ for every $x \in X$) and non-degenerate in the ordinary (non-equivariant) sense (i.e. embeds ordinary simplexes) is also equivariantly nondegenerate. In the ordinary case, i.e., without the presence of any group action, the assumption $\dim X \leq \dim M$ is enough to guarantee that a simplicial map, after introducing subdivisions, can be approximated by a non-degenerate map. In the equivariant case a condition of the form $\dim X^H \leq \dim M^H$ for every subgroup H of G is not enough as simple examples show. Our basic Lemma for the equivariant case is Lemma 1 below. (We denote below $X^{>H} = \{x \in X \mid H \subsetneq G_x\}$.) From now on X will always denote a *compact* equivariant simplicial complex.

LEMMA 1. *Let* Y *and* Z *be equivariant subcomplexes of* X *such that* $Y \cap Z$ *is strongly full in* Y. *Let* $f:X \to \mathring{D}^n(\rho)$ *be a linear G-map such that* $f|:Y \cap Z \to \mathring{D}^n(\rho)$ *is isovariant and nondegenerate. Assume that*

$$\dim(Y^H - (Y^{>H} \cup Z)) \leq \dim D^n(\rho)^H - \dim D^n(\rho)^{>H} - 1$$

for every subgroup H *of* G. *Let* $\varepsilon > 0$ *be given. Then there exists a linear G-map* $h:X \to \mathring{D}^n(\rho)$ *such that* $h|:Y \to \mathring{D}^n(\rho)$ *is isovariant and nondegenerate and such that* h *is equivariantly* ε*-homotopic* rel$|Z|$ *to* f.

If we assume that f is a linear *isovariant* map then the situation is quite different and the dimension assumptions are of the type given in Lemma 3. Let us first state the following.

LEMMA 2. *Let* $f:X \to R^n(\rho)$ *be a linear isovariant map. Then there exists* $\delta > 0$ *such that any linear G-map* $h:X \to R^n(\rho)$ *satisfying* $d(f(x),h(x)) < \delta$, *for every* $x \in X$, *is isovariant and moreover isovariantly* δ*-homotopic to* f.

LEMMA 3. *Let* Y *and* Z *be equivariant subcomplexes of* X *such that* $Y \cap Z$ *is strongly full in* Y. *Let* $f: X \to \overset{o}{D}{}^n(\rho)$ *be a linear isovariant map such that* $f|: Y \cap Z \to \overset{o}{D}{}^n(\rho)$ *is nondegenerate. Assume that*

$$\dim(Y^H - (Y^{>H} \cup Z)) \leq \dim D^n(\rho)^H$$

for every subgroup H *of* G. *Let* $\epsilon > 0$ *be given. Then there exists a linear isovariant map* $h: X \to \overset{o}{D}{}^n(\rho)$ *such that* $h|: Y \to \overset{o}{D}{}^n(\rho)$ *is nondegenerate and such that* h *is isovariantly* ϵ*-homotopic* rel$|Z|$ *to* f.

Using the above lemmas one can now prove Theorems 4 and 5. The proof of this step is very similar to the proof of the corresponding step in the ordinary non-equivariant case. (Compare with the proof of Lemma 7.2 in Hudson [2] or the proof of Theorem 1.6.10. Part 1 in Rushing [3].) If $f: X \to M$ is a G-map and X^H_α denotes some component of X^H we denote by $M^H_{f(\alpha)}$ the component of M^H containing $f(X^H_\alpha)$.

THEOREM 4. *Let* Y *be an equivariant subcomplex of* X *and let* M *be an equivariant combinatorial manifold. Let* $f: X \to M$ *be a simplicial G-map such that* $f|: Y \to M$ *is equivariantly nondegenerate. Assume that*

$$\dim(X^H_\alpha - (X^{>H}_\alpha \cup Y)) \leq \dim M^H_{f(\alpha)} - \dim M^{>H}_{f(\alpha)} - 1$$

for every subgroup H *of* G *and each component* X^H_α *of* X^H. *Let* $\epsilon > 0$ *be given. Then there exist equivariant subdivisions* X′ *of* X *and* M′ *of* M *and an equivariantly nondegenerate simplicial G-map* $h: X' \to M'$ *such that* h *is equivariantly* ϵ*-homotopic* rel$|Y|$ *to* f.

THEOREM 5. *Let* X, Y *and* M *be as above. Let* $f: X \to Y$ *be a simplicial isovariant map such that* $f|: Y \to M$ *is equivariantly nondegenerate. Assume that*

$$\dim(X_\alpha^H - (X_\alpha^{>H} \cup Y)) \le \dim M_{f(\alpha)}^H$$

for every subgroup H *of* G *and each component* X_α^H *of* X^H. *Let* $\varepsilon > 0$ *be given. Then there exist equivariant subdivisions* X′ *of* X *and* M′ *of* M *and an equivariantly nondegenerate simplicial G-map* $h: X' \to M'$ *such that* h *is isovariantly* ε*-homotopic* rel $|Y|$ *to* f.

REFERENCES

1. G. Bredon. *Introduction to compact transformation groups.* Academic Press. New York and London, 1972.

2. J.F. Hudson. *Piecewise-linear topology.* Benjamin, New York, 1969.

3. T.B. Rushing. *Topological embeddings.* Academic Press. New York and London, 1973.

UNIVERSITY OF HELSINKI

HELSINKI 10, FINLAND.

EQUIVARIANT RIEMANN-ROCH TYPE THEOREMS AND RELATED TOPICS

KATSUO KAWAKUBO

INTRODUCTION

Let G be a compact Lie group and let $h_G(\ )$ be an equivariant multiplicative cohomology theory. Let M and N be closed G-manifolds of class C^2. Then for a G-map $f: M \to N$, we shall define an 'equivariant Gysin homomorphism'

$$f_!: h_G(M) \to h_G(N)$$

under certain conditions.

In this talk, I would like to show that the study of our equivariant Gysin homomorphism is effective for various kinds of studies of transformation groups.

1. DEFINITION OF AN EQUIVARIANT GYSIN HOMOMORPHISM

In the non-equivariant case, a manifold has a unique stable normal bundle, which enables us to define a unique Gysin homomorphism. However, in the equivariant case, a stable normal bundle is not unique in a similar sense. Hence we need a device to get a uniqueness of our equivariant Gysin-homomorphism which is essential for our later uses.

Fix a property P, which will be given concretely later in respective category. For example, P = orientation in the oriented category. Denote by $\mathfrak{B}$ a set of G-vector bundles over finite CW-complexes and by B the set consisting of each element of $\mathfrak{B}$ together with the property P. Assume that the set B is endowed with the following properties:

1) Let $\xi \to X$ be an element of B and $f: Y \to X$ be a G-map where Y is a finite CW-complex. Then the induced bundle $f^*\xi$ belongs to $\mathfrak{B}$ and has a unique property P such that the bundle map preserves the property P.

2) B has a multiplication $B \times B \to B$ which induces the ordinary multiplication of $\mathfrak{B}$.

For a G-vector bundle $\xi \to X$, we denote by $D(\xi)$ (resp. $S(\xi)$) the total space of the disk bundle (resp. the sphere bundle) associated with ξ. An element $t_G(\xi)$ of $h_G(D(\xi), S(\xi))$ is called a Thom class (or h_G-orientation class) if for any compact G-invariant subspace Y of X, the correspondence $x \to t_G(\xi|Y)\, x$ gives an isomorphism

$$h_G(Y) \to h_G(D(\xi|Y) \ , \ \ S(\xi|Y)) \ .$$

We assume that we can assign to each element ξ of B a Thom class $t_G(\xi)$ such that

i) for a G-bundle map $f: \xi \to \xi'$ preserving P,

$$t_G(\xi) = f^* t_G(\xi') \ ,$$

ii) $t_G(\xi \times \eta) = t_G(\xi) \times t_G(\eta)$ for $\xi, \ \eta \in B$.

When such a correspondence is given, we say that B is oriented for h_G.

Now we are ready to define our equivariant Gysin homomorphism. Fix a property P and a set B with the property P. Let h_G be an equivariant multiplicative cohomology theory such that B is oriented for h_G. We consider a category of G-actions such that the G-vector bundles appearing in the category belong to the set B. (This meaning will become clear later.) Let M and N be h_G-oriented closed G-manifolds of class C^2. That is to say the tangent G-vector bundles of M and N belong to B, hence they are h_G-oriented. Then for a G-map $f:M \to N$, we define our equivariant Gysin homomorphism $f_!:h_G(M) \to h_G(N)$ as follows. As is well-known, there exists a G-representation space V and a G-embedding $e:M \to V$. Regarding V as a G-vector bundle over one point, we assume that V belongs to the set B. Then $f \times e:M \to N \times V$ is also a G-embedding. Denote by ν the normal G-vector bundle of the embedding $f \times e$. Then our equivariant Gysin homomorphism is defined by the composition of the following three homomorphisms which will be explained in a moment:

$$h_G(M) \xrightarrow{\phi_1} \tilde{h}_G(D(\nu)/S(\nu)) \xrightarrow{\phi_2} \tilde{h}_G(N\times D(V)/N\times S(V)) \xrightarrow{\phi_3} h_G(N)$$

EXPLANATION. $\tilde{h}_G$ denotes the reduced cohomology ring. Let

$$t_G(M) \in \tilde{h}_G(D(TM)/S(TM))$$

$$t_G(N) \in \tilde{h}_G(D(TN)/S(TN))$$

$$t_G(V) \in \tilde{h}_G(D(TV)/S(TV))$$

be the orientation classes respectively where TM, TN, TV denote the tangent G-vector bundles of M, N and V respectively. It is easy to see that we can choose a canonical orientation class $t_G(\nu)$ of ν such that

$$t_G(M) \times t_G(\nu) = (f \times e)^*(t_G(N) \times t_G(V)) .$$

Then the homomorphism ϕ_1 is defined to be the Thom isomorphism by making use of the Thom class $t_G(\nu)$. The homomorphism ϕ_2 is the induced homomorphism c^* by the natural collapsing map:

$$c: N \times D(V)/N \times S(V) \to D(\nu)/S(\nu) .$$

The homomorphism ϕ_3 is again defined by the Thom isomorphism in the manner of the definition of ϕ_1.

DEFINITION. When $f: M \to pt$, $f_!$ is called an index homomorphism and is denoted by Ind.

PROPOSITION. <u>The equivariant Gysin homomorphism is independent of all choices made and has the following properties:</u>

i) $f_!$ <u>depends only on the G-homotopy class of</u> f

ii) $f_!$ <u>is an</u> $h_G(pt)$-<u>module homomorphism</u>

iii) $(fg)_! = f_! g_!$

iv) $f_!(f^*(y).x) = y \cdot f_!(x)$ <u>for</u> $x \in h_G(M)$, $y \in h_G(N)$

v) <u>if</u> f <u>is a G-embedding of class</u> C^2 <u>with a normal bundle</u> $\nu (\in B)$, <u>then</u> $f^* . f_!(x) = \chi_G(\nu) \cdot x$ <u>for</u> $x \in h_G(M)$ <u>where</u> $\chi_G(\nu)$ <u>denotes the equivariant Euler class of</u> ν.

2. LOCALIZATION

Let S be the subset of $h_G(pt)$ consisting of Euler classes of the G-vector spaces $V \in B$ such that the group G acts on V without trivial direct summand. Here we regard a G-vector space as a G-vector bundle over one point. Then S is a multiplicative subset of $h_G(pt)$. It follows from ii) of the Proposition in §1 that we get a localized equivariant Gysin homomorphism: $S^{-1}f_!: S^{-1}h_G(M) \to S^{-1}h_G(N)$.

Denote by F_μ a component of the fixed point set of the

G-manifold M and by $i_\mu : F_\mu \to M$ the inclusion map and by N_μ the normal bundle of F_μ in M. There exists a G-vector space V without trivial direct summand and a G-map

$$f : M - \bigcup_\mu F_\mu \to V - \{0\} .$$

ASSUMPTION. The set B includes V and each N_μ can be assigned the property P so that N_μ belongs to B and $\chi_G(N_\mu)$ is a unit of $S^{-1}h_G(F_\mu)$.

We orient F_μ so that the orientation of the bundle N_μ followed by that of F_μ yields the restriction of the orientation of M. Then we have

THEOREM. Under the above assumption, the following diagram commutes:

$$\begin{array}{ccc} S^{-1}h_G(M) & \xrightarrow{\;S^{-1}\mathrm{Ind}\;} & S^{-1}h_G(pt) \\ \Big\downarrow \sum_\mu \dfrac{S^{-1}i_\mu^*}{\chi_G(N_\mu)} & \nearrow{\scriptstyle S^{-1}\mathrm{Ind}} & \\ \sum_\mu S^{-1}h_G(F_\mu) & & \end{array}$$

This is a generalization of the Atiyah-Bott-Segal Lefschetz fixed point Theorem and follows essentially from the uniqueness and the functorial properties of our equivariant Gysin homomorphism. Thus we get many equations between invariants of M and invariants of the fixed point set.

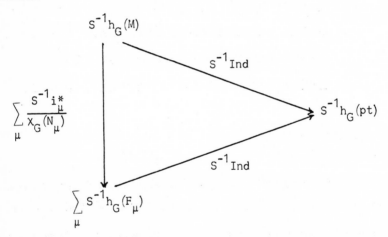

3. EQUIVARIANT RIEMANN-ROCH THEOREM IN Spin^c CATEGORY.

In the non-equivariant case, Dyer studied Riemann-Roch type theorems in general [4]. Unfortunately his arguments do not hold in the equivariant case. One of the reasons is that we cannot assume the compatibility of a multiplicative transformation with equivariant suspensions. On the contrary, the non-compatibility reflects the aspects of actions rather well.

Let us consider the case of spin^c. Let

$$G \to EG \to BG$$

be the universal principal G-bundle. We now use the following cohomology theory:

$$h_G(M) = H^*(EG \times_G M)$$

where $H^*(\ )$ is ordinary cohomology theory. In this category, the property P is ordinary orientation and B consists of oriented G-vector bundles over finite CW-complexes. Then B has the required properties 1, 2. We assign to each element $\xi \to X$ of B the usual Thom class $t_G(\xi)$ of the bundle $EG \times_G \xi \to EG \times_G X$. This assignment satisfies the required conditions. Thus B is oriented for $H^*(EG \times_G M)$ theory and we get our equivariant Gysin homomorphism

$$f_!: H^*(EG \times_G M) \to H^*(EG \times_G N)$$

for a G-map $f: M \to N$ where M and N are oriented G-manifolds of class C^2.

As usual a map $f: M \to N$ is called a c_1-map if we are given an element $c_1 \in H^2(M;Z)$ such that

$$c_1 \equiv W_2(M) - f^*W_2(N) \mod 2$$

where $W_2(\ )$ denotes the second Stiefel-Whitney class. We assume that $\dim M \equiv \dim N \bmod 2$. Recall the G-embedding $f \times e: M \to N \times V$ with normal bundle ν. Denote by Q the principal SO(2k)-bundle associated with ν. The action on ν induces an action on Q. As is well-known, if f is a c_1-map, then Q has the $\mathrm{spin}^c(2k)$ reduction $\tilde{Q}$ corresponding to c_1. Suppose that the action on Q lifts to an action on $\tilde{Q}$. For this, we can construct an obstruction theory. Then we have

THEOREM. *For $\xi \in K_G(M)$, there exists $\eta \in K_G(N)$ such that*

$$f_!(e^{c_G/2} \cdot ch_G(\xi) \cdot \hat{\mathfrak{a}}_G(M)) = ch_G(\eta) \cdot \hat{\mathfrak{a}}_G(N)$$

where c_G is the first Chern class of the complex line bundle

$$EG \times_G \tilde{Q} \times_{\mathrm{spin}^c(2k)} C \to EG \times_G M$$

and

$$ch_G(\xi) = ch(EG \times_G \xi), \quad ch_G(\eta) = ch(EG \times_G \eta)$$

$$\hat{\mathfrak{a}}_G(M) = \hat{\mathfrak{a}}(EG \times_G TM), \quad \hat{\mathfrak{a}}_G(N) = \hat{\mathfrak{a}}(EG \times_G TN).$$

REMARK. The correspondence $\xi \to \eta$ is not functorial in general. This contrasts with the non-equivariant case [1]. The reason is that the correspondence depends on the liftings and we cannot choose a canonical lifting in general so that the correspondence is functorial. In fact there are many examples showing that the correspondence depends on the liftings.

As in the non-equivariant case, there are a lot of applications of the theorem. For example, we shall have integrality and divisibility theorems of equivariant characteristic numbers. On the other hand there are particularly interesting applications peculiar to equivariant cases. Namely, by combining equivariant Riemann-Roch type theorems and the localization theorem in §2, we can often conclude that some global invariants are non-equivariant or even zero. This is in fact my motivation to establish the equivariant Riemann-Roch type theorems. Thus we have

COROLLARY 1. *Let* M *be a connected, spin, differentiable manifold of class* C^2 *with non-trivial* S^1*-action. Then we have*

$$\hat{\mathfrak{a}}(M)[M] = 0 .$$

REMARK. This gives a topological proof of a generalization of [2], which answers a problem of I.M. Singer [5].

COROLLARY 2. *Let* M *be a* spin^c*-manifold of class* C^2 *and* $\xi_i \to M$ $(i = 1$ *or* $2)$ *be complex vector bundles. Then there exists an element* c *of* $H^2(CP(\xi_1 \oplus \xi_2);Z)$ *such that*

$$e^{c/2}\hat{\mathfrak{a}}(CP(\xi_1 \oplus \xi_2))[CP(\xi_1 \oplus \xi_2)] = 0$$

where $CP(\xi_1 \oplus \xi_2)$ *denotes the total space of the complex projective space bundle associated with* $\xi_1 \oplus \xi_2$.

4. G-HOMOTOPY TYPE INVARIANCE OF EQUIVARIANT STIEFEL-WHITNEY CLASSES

In this section we consider the non-oriented category and make use of $H^*(EG \times_G M;Z/2)$-theory. In this case property $P = \emptyset$ and B is the set of (non-oriented) G-vector bundles over finite CW-complexes. We assign to each element $\xi \to X$ of B the unique Thom class

$$t_G(\xi) \in H^*(EG \times_G (D(\xi), S(\xi)); Z/2)$$

of the vector bundle $EG \times_G \xi \to EG \times_G X$. Then these satisfy the conditions in §1.

Denote by S_q^i the Steenrod i-th squaring operation. Put

$$S_q = \sum_{i=0}^{\infty} S_q^i \quad \text{and} \quad S_q^{-1} = \frac{1}{1+(S_q-1)}$$

as formal power series. For a G-vector bundle $\xi \to X$, we put

$$W_G(\xi) = \phi^{-1} S_q \phi(1), \qquad V_G(\xi) = S_q^{-1} \phi^{-1} S_q \phi(1)$$

$$\overline{W}_G(\xi) = S_q \phi^{-1} S_q^{-1} \phi(1), \qquad \overline{V}_G(\xi) = \phi^{-1} S_q^{-1} \phi(1)$$

where $\phi: H^*(EG \times_G M; Z_2) \to H^*(EG \times_G (D(\xi), S(\xi)); Z/2)$ is the Thom isomorphism. Then $W_G(\xi)$ is nothing but the total Stiefel-Whitney class of the vector bundle $EG \times_G \xi \to EG \times_G X$ and is called the equivariant total Stiefel-Whitney class of ξ. Then we have

THEOREM. <u>For a G-map</u> $f: M \to N$, <u>we have</u>

$$f_!(x \cdot V_G(M)) = (S_q^{-1} f_!(S_q x)) \cdot V_G(n)$$

$$f_!(x \cdot \overline{W}_G(M)) = (S_q f_!(S_q^{-1} x)) \cdot \overline{W}_G(n)$$

<u>for</u> $x \in H^*(EG \times_G M; Z/2)$.

As is well-known, EG can be written as

$$EG = \lim EG^n$$

where EG^n is an n-connected free G-manifold. By making use of EG^n instead of EG, we can define $f_!^n: H^*(EG^n \times_G M; Z/2) \to H^*(EG^n \times_G N; Z/2)$ and $W_G^n(\xi)$, $V_G^n(\xi)$, $\overline{W}_G^n(\xi)$, $\overline{V}_G^n(\xi)$ similarly and we have

THEOREM n. (n=1,2,...). <u>For a G-map</u> $f: M \to N$, <u>we have</u>

$$f_!^n(x \cdot V_G^n(M)) = (S_q^{-1} f_!^n(S_q x)) \cdot V_G^n(N)$$

$$f_!^n(x\cdot\overline{W}_G^n(M)) = (S_q f_!^n(S_q^{-1}x))\cdot\overline{W}_G^n(N)$$

for $x \in H^*(EG^n \times {}_G M;Z/2)$.

This Theorem is used to prove the following

THEOREM. $W_G(M)$, $V_G(M)$, $\overline{W}_G(M)$ _and_ $\overline{V}_G(M)$ _are all invariants of the G-homotopy type._

COROLLARY. $(Z/2)^k$-_homotopy equivalent manifolds are_ $(Z/2)^k$-_bordant._

This follows from the fact that non-oriented $(Z/2)^k$-bordism classes are characterized by equivariant Stiefel-Whitney numbers due to tom Dieck [3].

5. RELATED TOPICS

Along our line, we could develop our theory as follows.

(1) We shall have equivariant Riemann-Roch type theorems in various categories.

(2) Concerning Landweber-Nowikov operations, we shall also have equivariant Wu type formulae.

(3) By making use of the equivariant Gysin homomorphism, we shall define equivariant characteristic numbers. Equivariant characteristic numbers are used to characterize bordism classes of G-manifolds. For example, bordism classes of oriented toral manifolds are characterized by the equivariant characteristic numbers modulo 2 torsions.

(4) We shall have equivariant non-embedding theorems. So far, equivariant non-embedding theorems have been studied by Liulevicius, Bix and Stong in the case of involutions. Our Theorem

is applicable for any compact Lie group action.

(5) We shall have vanishing theorems of some genera in general.

REFERENCES

1. M.F. Atiyah, R. Bott and A. Shapiro, Clifford modules, Topology 3 (1964), Suppl. 1, 3-38.

2. M.F. Atiyah and F. Hirzebruch, Spin manifolds and group actions, Essays on Topology and Related Topics Mémoirs dédiés à Georges de Rham, pp 17-28, Springer, Berlin and New York, 1970.

3. T. tom Dieck, Characteristic numbers of G-manifolds I, Invent. Math. 13 (1971) 213-224.

4. E. Dyer, Cohomology Theories, Benjamin, 1969.

5. I.M. Singer, Recent applications of index theory for elliptic operators, Proc. Symposia Pure Math., Amer. Math. Soc. 23(1973), 11-31.

UNIVERSITY OF BONN

BONN, WEST GERMANY

and

UNIVERSITY OF OSAKA

OSAKA, JAPAN

KNOTS AND DIFFEOMORPHISMS

M. KRECK

In this note I will announce some results concerning the connection between bordism of knots and diffeomorphisms and state some problems.

Consider the bordism group of knots C_k consisting of bordism classes of knots $\Sigma^k \hookrightarrow S^{k+2}$, Σ^k a k-dimensional homotopy sphere. Kervaire has shown that this group is zero for k even [2]. Levine has proved that for $k \geq 3$ there is an embedding $C_{2k-1} \to W_{(-1)^k}(Z,Q)$, the Witt group of $(-1)^k$-symmetric isometric structures over Q ([6], for definition of $W_{(-1)^k}(Z,Q)$ compare [7]). Kervaire has shown that $W_{(-1)^k}(Z,Q)$ is of the form $Z^\infty \oplus (Z/4)^\infty \oplus (Z/2)^\infty$. Several people have stated that C_{2k-1} is of the form $Z^\infty \oplus (Z/4)^\infty \oplus (Z/2)^\infty$, too, but no proof has appeared. It is rather easy to see that this holds for k odd and that for k even $C_{2k-1} \otimes Q$ is Q^∞ and C_{2k-1} contains infinitely many torsion elements [7].

Another group which is classified in terms of isometric structures is the bordism group of n-dimensional orientation preserving diffeomorphisms Δ_n. In [3] I have shown that this group for n odd $(n \neq 3)$ is classified in terms of the manifold and the mapping torus of the diffeomorphism. For n even $(n > 2)$ we need another invariant given by the isometric structure of a diffeomorphism which lies in $W_{(-1)^k}(Z,Z)$, the Witt group of $(-1)^k$-symmetric isometric structures over Z ([4], [5]). $W_{(-1)^k}(Z,Z)$ is a subgroup of $W_{(-1)^k}(Z,Q)$ and is of the form $Z^\infty \oplus (Z/4)^\infty \oplus (Z/2)^\infty$, too.

The description of bordism groups of knots and diffeomorphisms gives an algebraic connection between these groups and the question is whether there is any geometrical connection. This can be obtained by considering fibred knots. A fibred knot is a knot $\Sigma^k \hookrightarrow S^{k+2}$ together with a fibration of the complement of the knot over S^1. Examples of fibred knots are the Brieskorn spheres. The subgroup of C_k represented by fibred knots is denoted by C_k^F.

Now, for a fibred knot $\Sigma^k \hookrightarrow S^{k+2}$ we can construct a diffeomorphism as follows. Let $g: W \to W$ be the diffeomorphism classifying the fibration $S^{k+2} - \Sigma^k \to S^1$. The fibre W can be considered as a manifold with boundary Σ and $g|_{\partial W}$ is the identity. To obtain a diffeomorphism on a closed manifold we consider the double $W \cup_\Sigma (-W)$ and the diffeomorphism $g \cup \mathrm{Id}$.

THEOREM 1: <u>For</u> $k \geq 3$ <u>this construction induces a homomorphism</u>

$$\rho: C_{2k-1}^F \to \Delta_{2k}$$

<u>This homomorphism is injective</u>.

The proof is given by comparison of the invariant classifying knots and diffeomorphisms, so it is not a direct proof. It would be interesting to have such a proof. This could give an idea of how to prove Theorem 1 for low dimensions. We state this as a problem.

PROBLEM. Give a direct geometrical proof of Theorem 1 and extend it to low dimensions.

As it stands Theorem 1 is not of much worth as we have not yet said anything about C^F_{2k-1}.

THEOREM 2. $C^F_{2k-1} \otimes Q$ *is isomorphic to* Q^∞. *This fact is detected by Brieskorn spheres*.

PROBLEM. Determine the torsion subgroup of C^F_{2k-1}.

Now we want to state some problems concerning low dimensions. In [1] Casson and Gordon have shown that the bordism classes of classical knots $S^1 \hookrightarrow S^3$ are not determined by their isometric structure. This fact and the connection between knots and diffeomorphisms dare to raise the following conjecture.

CONJECTURE. Bordism classes of diffeomorphisms on surfaces are not determined by their isometric structure.

REMARK. A consequence of this conjecture would be that at least a relative version of the h-cobordism Theorem does not hold in dimension 4.

REFERENCES

1. A.J. Casson, C. McA. Gordon: Cobordism of classical knots. Preprint 1975.

2. M. Kervaire: Les noeuds de dimensions supérieures. *Bull. Soc. Math. France* 93 (1965), 225-271.

3. M. Kreck: Cobordism of odd-dimensional diffeomorphisms, to appear in *Topology* (1976).

4. M. Kreck: Bordism of diffeomorphisms, to appear in *Bull. A.M.S.* (1976).

5. M. Kreck: Bordismusgruppen von Diffeomorphismen. Preprint Bonn (1976).

6. J. Levine: Knot cobordism groups in codimension two. Comm. Math. Helv. 44 (1969), 229-244.

7. W.D. Neumann: Equivariant Witt rings. Preprint. Bonn (1976).

UNIVERSITY OF BONN

BONN, WEST GERMANY

SOME REMARKS ON FREE DIFFERENTIABLE INVOLUTIONS ON HOMOTOPY SPHERES

PETER LÖFFLER

Let G denote the group Z/2. Let $R^{n,m}$ denote the n+m-dimensional real vector space with a non-trivial G-action on the first n coordinates. We call a G-manifold (n,m)-framed, if $tM \oplus \epsilon^{k,\ell} = \epsilon^{k+n,\ell+m}$. Denote the corresponding bordism group by $\pi^G_{n,m}$ and by $\pi^G_{n,m}[1]$ if no fixed points are allowed. Let us denote by $S^{n,m}$ the sphere in $R^{n,m}$ in some equivariant metric and introduce $\omega^G_{n,m} = \lim_{k,\ell} [S^{n+k,m+\ell}, S^{k,\ell}]^0_G$, where $[\ ,\ ]^0_G$ denotes equivariant basepoint preserving homotopy classes of maps, and $\omega^G_{n,m}[1] = \lim_{k,\ell}[S^{n+k,m+\ell}, S^{k,\ell} \wedge EG^+]^0_G$. A well-known transversality argument assures that the Pontryagin-Thom-construction yields an isomorphism $\pi^G_{n,m} = \omega^G_{n,m}$ and $\pi^G_{n,m}[1] = \omega^G_{n,m}[1]$ at least for m = -1.

Now if Σ^n is an n-dimensional homotopy sphere with a free differentiable involution T then an easy lemma asserts that (Σ^n, T) is (n+1,-1)-framed.

If we denote by θ^G_n $({}^{fr}\theta^G_n)$ the diffeomorphism classes of n-dimensional homotopy spheres with free differentiable involution (together with a specific framing) we get a map of sets ${}^{fr}\theta^G_{n-1} \to \omega^G_{n,-1}[1]$. One can describe the image by surgery techniques and the existence of a degree one mapping.

If we denote by im $J^G_{n,-1}$ the subgroup of $\omega^G_{n,-1}$ generated by the standard sphere with different framings (we have im $J^G_{n,-1} = 0$

for $n \not\equiv 0(4)$) we can define $\Theta^G_{n-1} \to \hat{\omega}^G_{n,-1} = \omega^G_{n,-1}/\mathrm{im}\, J^G_{n,-1}$.

PROPOSITION 1. $\Theta^G_{n-1} \to \hat{\omega}^G_{n,-1}$ *is surjective.*

Now Bredon and Landweber (Annals of Math. 89, 1969) have computed the kernel of $\omega^G_{n,-1}[1] \to \omega^G_{n,-1}$. It is a cyclic subgroup Z/b_n with $b_n = 2a_n$ $(4a_n)$ with a_n = order of $\widetilde{KO}(RP(n-1))$ if $n \not\equiv 0(4)$ $(n \equiv 0(4))$.

PROPOSITION 2. $x \in Z/b_n$ *represents a homotopy sphere if*

x *odd for* $n \equiv 2(4)$

$x \equiv \pm 1(8)$ *for* $n \not\equiv 2(4)$

As Hoo and Mahowald have computed $\omega^G_{n,-1}[1]$ up to $n \leq 14$ this result gives a classification of free involutions on homotopy spheres up to this dimension except for $n = 9, 10$.

If we denote by Σ_d the Brieskorn sphere of dimension $4n + 1$ defined by $z_0^d + z_1^2 + \ldots + z_{2n+1}^2 = 0$, d odd, and involution $(z_0, z_1, z_2, \ldots) \to (z_0, -z_1, -z_2, \ldots)$ then we have

PROPOSITION 3. Σ_d *is diffeomorphic to* $\Sigma_{d'}$, *up to an action of* $L_2(Z/2,+)$, *if and only if* $d \equiv d'$ (2^{2n+2}).

Let $e: \omega^G_{n,-1} \to \omega^G_{n-1,-1}$ be the Smith-homomorphism and let $\mathcal{L} = \varprojlim(\omega^G_{n,-1}, e)$. Then a recent result of Snaith says that $\mathcal{L} \neq 0$.

If we define a standard involution to be a free involution on a homotopy sphere such that this element bounds in $\omega^G_{n,-1}$ then Proposition 2 gives a classification of the standard involutions (at least for $n \not\equiv 0(4)$). Now $\mathcal{L} \neq 0$ says

PROPOSITION 4. *There exist non-standard involutions on the standard sphere* S^n *in almost all dimensions.*

UNIVERSITY OF GÖTTINGEN, GÖTTINGEN, WEST GERMANY

COMPACT TRANSITIVE ISOMETRY GROUPS

GORDON W. LUKESH

In this note we would like to announce recent results concerning homogeneous Riemannian manifolds. The proofs will appear elsewhere. Theorem 1 is a classification of compact homogeneous Riemannian manifolds having 'large' isometry groups. Theorem 2 is concerned with a conjecture and theorem due to Wu-yi Hsiang on the degree of symmetry of homogeneous manifolds. In particular, we show that a result of his ([1]) concerning the second Stiefel manifold $V^{n,2} = SO(n)/SO(n-2)$ is in error.

This work was completed while the author was a graduate student at the University of Massachusetts, under the direction of Professor Larry Mann. The author would like to thank Professor Mann for suggesting these problems, and for many useful conversations.

THEOREM 1. Let $M = K/H$ be an m-dimensional compact homogeneous Riemannian manifold with K an effective Lie group of isometries of dimension at least $m^2/4 + m/2$, $(m \geq 19)$. Then one of the following must hold:

1. If M is simply connected, then either

   (i) $M = CP^n$ $(m=2n)$. The metric on M is determined up to a scale factor, and K is locally isomorphic to $SU(n+1)$.

or (ii) $M = S^k \times V^{m-k}$, $k \geq m/2$. S^k is isometric to a standard sphere, $V^{m-k} = K_2/H_2$ is a simply connected Riemannian homogeneous *manifold, and the metric on* M *is the product metric.* K *is locally isomorphic to* Spin(k+1) × K_2 *where* dim $K_2 \le (m-k)(m-k+1)/2$.

or (iii) $M = S^m = U(n)/U(n-1)$ ($m=2n-1$). *There are uncountably many homothetically distinct homogeneous metrics on* M *having* U(n) *as full isometry group.*

2. *If* M *is not simply connected, then either*

(iv) $M = CP^n \times S^1$ ($m=2n+1$). *The metric on* M *is the product metric of a metric on* CP^n *(determined up to a scale factor) and a usual metric in* S^1. K *is locally isomorphic to* U(n+1).

or (v) $M = S^k \times_A V^{m-k}$ *where* A *is a group of order two acting freely on* $S^k \times V^{m-k}$, $k \ge m/2$. S^k *is isometric to a standard sphere and the metric on* M *is locally a product metric.* K *is as in part* (ii).

or (vi) M *is a simple lens space finitely covered by* $S^m = U(n)/U(n-1)$. *Such a lens space possesses uncountably many homothetically distinct homogeneous metrics having* U(n) *as full isometry group.*

or (vii) $M = RP^k \times V^{m-k}$, $k \ge m/2$. RP^k *is isometric to a standard real projective space,* V^{m-k} *is Riemannian homogeneous, and the metric on* M *is the product metric.* K *is as in part* (ii).

REMARKS. The proof relies on two basic facts. First, the isometry group of a compact Riemannian manifold is compact, thus we may apply results from compact transformation groups, most notably, Mann [4]. Second, invariant metrics on $M = K/H$ are in one-one correspondence with inner products on the tangent space to the coset H in M, invariant by the linear isotropy action of H. For example, if the linear action of H is irreducible, the invariant metrics on M are determined up to a scale factor. The proof of this theorem will appear in [2].

If the linear isotropy action is reducible, as in (iii), we have shown that it is possible to vary a given homogeneous metric through a family of homogeneous metrics. By computing the curvature tensor and sectional curvatures for these metrics, it is possible to distinguish the resulting Riemannian manifolds. In (iii) all but one of the spheres obtained has non-constant curvature. For details, see [3].

Another application of this 'variation' technique involves a concept due to Hsiang: Define the degree of symmetry of a differentiable manifold M to be the maximal dimension of all isometry groups of all possible Riemannian metrics on M. If $M = K/H$, where K is compact, semi-simple, then there is a 'natural' metric on M, associated with the Cartan-Killing form of the Lie algebra of K.

CONJECTURE. (Hsiang, [5]). The natural metric on a homogeneous manifold is the most symmetric metric.

In [1], Hsiang tested this conjecture with the second Stiefel manifold $V^{n,2} = SO(n)/SO(n-2)$, which has $SO(n) \times SO(2)$ as full compact transitive diffeomorphism group. He claims that the natural metric on $V^{n,2}$ ($n \geq 31$, odd) alone, up to a scale factor, has $SO(n) \times SO(2)$ as isometry group, and that all other metrics have isometry groups of strictly lower dimension. However, using the variation technique we have shown the following:

THEOREM 2. The second Stiefel manifold $V^{n,2}$ ($n \geq 31$, odd) has uncountably many homothetically distinct homogeneous metrics having $SO(n) \times SO(2)$ as isometry group.

Details will appear in [3].

REFERENCES

1. W.Y. Hsiang. The natural metric on SO(n)/SO(n-2) is the most symmetric metric. Bull. Amer. Math. Soc. 73 (1967), 55-58.

2. G. Lukesh. Compact homogeneous Riemannian manifolds, to appear.

3. G. Lukesh. Variations of metrics on homogeneous manifolds, to appear.

4. L.N. Mann. Highly symmetric homogeneous spaces. Canadian J. Maths. 26 (1974), 291-293.

5. P. Mostert, (Editor). Proc. Conf. on Transf. Groups. New Orleans, 1967 Berlin-Heidelberg-New York Springer, 1968.

UNIVERSITY OF MASSACHUSETTS
AMHERST, MA 01002, USA

Present address:

UNIVERSITY OF TEXAS
AUSTIN, TEXAS 78712, USA

A PROBLEM OF BREDON CONCERNING HOMOLOGY MANIFOLDS

W. J. R. MITCHELL

Let L denote a principal ideal domain and X a locally compact Hausdorff topological space of finite cohomological dimension over L. X is called a *generalised n-manifold over L* ($n\text{-gm}_L$) if

$$\text{(i)} \quad \forall\, x \in X\,, \quad H_*(X, X-x; L) = \begin{cases} L & \text{if} \quad * = n \\ 0 & \text{if} \quad * \neq n \end{cases}$$

(ii) X is clc_L .

[Here sheaf cohomology and Borel-Moore homology are understood; under our assumptions X is clc_L if and only if for U open and $x \in U$, there exists V with $x \in V \subset U$ and im $j^*: H^*_c(V) \to H^*_c(U)$ finitely generated.]

Bredon [1] calls a space X an $n\text{-hm}_L$ if it satisfies (i). These concepts arise in the study of group actions on manifolds. The clc_L condition (ii) is vital for many arguments, but can be troublesome to verify. In [2] Bredon asks if it is a consequence of (i). In the same paper he proves this when L is a countable field.

THEOREM. *Let* X *and* L *be as above. Assume further that* L *is countable,* X *is first countable and that for* U *open in* X, $H^*_c(U;L)$ *is countable. Then if the modules* $H_p(X, X - x; L)$ *are for all* $x \in X$ *and all integers* p, X is clc_L.

The proof, which will appear elsewhere, uses a considerable elaboration of the methods of [2]. The assumption on $H_c^*(\ )$ is probably unnecessary, and in any case holds if X is separable metric. The first countability assumption on X is essential to the method of proof, even if L is a field. Indeed, using a non-principal ultrafilter on the integers one easily constructs an inverse system of Z/2-modules $\{V_\alpha\}$ such that dir lim $\mathrm{Hom}(V_\alpha, Z/2) = Z/2$, but with all the bonding maps having infinite dimensional images. This is a counter-example to the obvious generalisation of lemma 2.1 of [2]. Note that 'change of rings' difficulties prevent a straightforward deduction from the results of [2] of the following Corollary to our Theorem.

COROLLARY. *With* X, L *as in the Theorem, assume* X *is an* $n\text{-hm}_L$. *Then* X *is an* $n\text{-gm}_L$.

REFERENCES

1. G.E. Bredon. *Sheaf Theory*. McGraw-Hill, 1967.
2. G.E. Bredon. Generalised Manifolds, Revisited, in *Topology of Manifolds* (Georgia Conference, 1969) ed. J.C. Cantrell & C.H. Edwards Jr., Markham 1970.

CHRIST'S COLLEGE

CAMBRIDGE